San Francisco

Big City Food Biographies Series

Series Editor
Ken Albala, University of the Pacific, kalbala@pacific.edu

Food helps define the cultural identity of cities in much the same way as the distinctive architecture and famous personalities. Great cities have one-of-a-kind food cultures, offering the essence of the multitudes who have immigrated there and shaped foodways through time. The **Big City Food Biographies** series focuses on those metropolises celebrated as culinary destinations, with their iconic dishes, ethnic neighborhoods, markets, restaurants, and chefs. Guidebooks to cities abound, but these are real biographies that will satisfy readers' desire to know the full food culture of a city. Each narrative volume, devoted to a different city, explains the history, the natural resources, and the people that make that city's food culture unique. Each biography also looks at the markets, historic restaurants, signature dishes, and great cookbooks that are part of the city's gastronomic makeup.

Books in the Series

New Orleans: A Food Biography, by Elizabeth M. Williams

San Francisco

A Food Biography

• Erica J. Peters

ROWMAN & LITTLEFIELD
Lanham • New York • Toronto • Plymouth, UK

Published by Rowman & Littlefield
4501 Forbes Boulevard, Suite 200, Lanham, Maryland 20706
www.rowman.com

10 Thornbury Road, Plymouth PL6 7PP, United Kingdom

British Library Cataloguing in Publication Information Available

Library of Congress Cataloging-in-Publication Data

Peters, Erica J.
  San Francisco : a food biography / Erica J. Peters.
    pages cm
  Includes bibliographical references and index.
  ISBN 978-0-7591-2151-5 (cloth : alk. paper) — ISBN 978-0-7591-2153-9 (ebook)
  1. Food habits—California—San Francisco Bay Area. 2. Food industry and trade—California—San Francisco Bay Area. 3. Cooking, American—California style. 4. Agriculture—California. I. Title.
  GT2853.U5P48 2013
  394.1'209794'6—dc23   2013018090

Contents

~

Series Foreword

Big City Food Biographies

Cities are rather like living organisms. There are nerve centers, circulatory systems, structures that hold them together, and, of course, conduits through which food enters and waste leaves the city. Each city also has its own unique personality, based mostly on the people who live there, but also the physical layout, the habits of interaction, and the places where people meet to eat and drink. More than any other factor, it seems that food is used to define the identity of so many cities. Simply say any of the following words and a particular place immediately leaps to mind: bagel, cheese steak, muffaletta, chowda, cioppino. Natives, of course, have many more associations—their favorite restaurants and markets, bakeries and doughnut shops, pizza parlors, and hot dog stands. Wherever you go, even the restaurants seem to have their unique vibe. Some cities boast great steakhouses or barbecue pits, others their ethnic enclaves and more elusive specialties such as Frito pie in Santa Fe, Cincinnati chili, and the Chicago deep-dish pizza. Tourists might find snippets of information about such hidden gems in guidebooks; the inveterate flaneur naturally seeks them out personally. For the rest of us, this is practically uncharted territory.

These urban food biographies are not meant to be guidebooks but rather real biographies, explaining the urban infrastructure, the natural resources that make each city unique, and, most importantly, the history, people, and neighborhoods. Each volume is meant to introduce you to the city, or reacquaint you with an old friend in ways you may never have considered. Each biography also looks at the historic restaurants, signature dishes, and great cookbooks that reflect each city's unique gastronomic makeup.

These food biographies also come at a crucial juncture in our culinary history as a people. Not only do chain restaurants and fast food threaten the existence of our gastronomic heritage, but we are increasingly mobile as a people, losing our deep connections to place and the cooking that happens in cities over the generations with a rooted population. Moreover, signature dishes associated with individual cities become popularized and bastardized and are often in danger of becoming caricatures of themselves. Ersatz versions of so many classics, catering to the lowest common denominator of taste, are now available throughout the country. Our gastronomic sensibilities are in danger of becoming entirely homogenized. The intent here is not, however, to simply stop the clock or make museum pieces of regional cuisines. Cooking must and will evolve, but understanding the history of each city's food will help us make better choices, will make us more discerning customers, and perhaps more respectful of the wonderful variety that exists across our great nation.

Ken Albala
University of the Pacific

~

Preface

My first intimate contact with California came through its food. In January 1987 I was seventeen and living in Ithaca, New York. That winter my father was visiting a mathematical institute in Berkeley. My mother had gone along with my two younger siblings, and she would call me with amazing stories of their discoveries. They ate mangos, kiwis, and papayas for breakfast—in January. What were these fruits, what did they taste like, how did you eat one? I was full of questions. She told me about a meal at Chez Panisse. We had been to Paris, and I thought I knew something about French restaurants—this one sounded different. One time she explained sushi to me. During another call she described her new favorite Thai restaurant: she had eaten a dish with tiny green chilis that nearly blew off the top of her head. While we spoke, I was sorry I skipped going with them to experience all these new tastes.

None of this resembled the stories my Brooklyn-raised mother had told of visiting San Francisco with her own parents in 1950. They had had tea at the St. Francis and eaten at Joe DiMaggio's restaurant. She remembered Fisherman's Wharf for its corn dogs; even in Coney Island she had never seen such a thing. In 1987 my parents had invited me to come with them, but the idea of corn dogs on Fisherman's Wharf and tea at the St. Francis did not entice me away from my high school friends. Mangos and sushi, on the other hand . . .

Four years later I moved to the San Francisco Bay Area, eager for new experiences culinary and otherwise. I fell in love with the region and slowly put down roots. The country was in recession, and San Francisco was still recovering from the 1989 earthquake. But food was central to people's lives here, and I was delighted to finally be able to try mangos. And Chez Panisse. And the Tadich Grill. And dim sum at Yank Sing. Over the years, I enjoyed breakfast at Dottie's True Blue Cafe, brunch at the Palace Hotel, lunch at Max's Opera Cafe, shopping at the Ferry Plaza Farmers Market, dinner at the Slanted Door, late nights at the Tonga Room, and many more culinary adventures. Everywhere I ate in San Francisco, I was surrounded by history along with delectable food. Food history was barely a field at that point, but I found it intriguing. Quite a bit later, I realized that I could actually write about the food history of San Francisco. This book has been a way for me to investigate all the wonderful food that San Franciscans ate before I got here to share the good times. Taking my arrival in California as the end point for this history is admittedly personal and idiosyncratic. But, then, the book itself is an idiosyncratic take on one city's cuisine—a pleasure to research and write.

Acknowledgments

Many people have helped me along the way. For warmly welcoming me to the Bay Area and sharing their insights with me, I thank Ken Albala, Naomi Andrews, Laura Martin Bacon, Ruth Begell, Robert Brower, Pam Elder, Jeannette Ferrary, Jeremy Fletcher, Nancy Freeman, Bert Gordon, Donna Green-Tye, Andy Griffin, Sheila Himmel, Dianne Jacob, Norma Kobzina, Mary Margaret Pack, Celia Sack, Andrew Sigal, Betty Teller, Thy Tran, Jessica Weiss, and Frankie Whitman. For rich discussions and comparative ideas about other cities, I am grateful to Daniel Block, Matthew Booker, Karen Carr, Marilyn Peters Dunn, Laresh Jayasanker, Rachel Laudan, Lisa Leff, Sari Locker, Michael McKernan, Sandy Oliver, Amy Ross, Leonard Schmieding, Andy Smith, and Liz Williams. The staff at the University of California's Bancroft Library and the San Francisco Public Library greatly facilitated my research, as did the Stanford University history department. At Rowman & Littlefield, Wendi Schnaufer, Karen Ackermann, and Andrea Kendrick provided invaluable help at every stage. My family has long supported my food history interests. I especially thank my parents, Steve and Lynn Lichtenbaum, for those tantalizing tales of San Francisco cuisine, and David, Charlotte, and Marc for much encouragement and love.

~

Introduction

San Francisco's location on the Pacific Ocean has always animated its history. The weather comes off the ocean; much of the seafood comes directly or indirectly from the coastal waters; and immigrants to San Francisco have long arrived through the strait known as the Golden Gate, even when their route took them around South America or across Panama and then up the West Coast. Along with the weather, products, and people, culinary ideas often come to California from farther west—from the "Far East," across the Pacific—rather than from the eastern United States. Chinese immigrants brought their expertise in stir-frying fresh vegetables with imported flavors; Japanese immigrants brought delicate dishes, from tea cakes to sushi. Immigrants from all over brought their specialties—ways to prepare fish, to barbecue, to make noodles and breads, to make soups and stews. In San Francisco everyone explored and experimented with each other's culinary traditions, emerging with a potpourri of popular favorites.

At every stage in the city's history, these diverse delights encountered Northern California's other culinary constant: the richness of the natural environment. Newcomers have always been amazed by the local land and water. The soil produces fruits and vegetables remarkable for their size and quality. Each generation of immigrant cooks found ways to incorporate the luscious local produce into their familiar recipes. The results help explain why different communities have been willing to try each other's foods—San Franciscans trusted that many magnificent dishes could be made from such mouthwatering ingredients.

But the diversity of dishes also raised new trust issues. Without a shared culinary background, the citizens of this cosmopolitan city faced uncertainty about what basic ingredients lay underneath the splendid sauces they commonly tasted in French, Chinese, or Mexican restaurants. Could those delicious concoctions mask unreliable components? San Franciscans found their solution in reputation. Chefs made names for themselves with the excellence of their provisions, many of them sourced from local farms and fishmongers. As far back as the Gold Rush era and up to the present, Bay Area restaurants and retailers have worked hard to build customer trust in their providers and in their reputation for reliability. The success of many a new San Francisco venture has initially depended on demonstrating a link, however tenuous, to a former establishment with a great reputation.

Ever since its early days, San Francisco's history has been intimately bound up with its changing foodways. Yerba Buena, a Spanish name for the settlement, meant "good herb" and came from the ubiquity in the area of a particular kind of wild mint. Local Native Americans helped early Spanish missionaries to Northern California survive, by sharing nuts, fruits, roots, fish, and game; later, these Ohlone peoples were pressed into service preparing wheat or corn for tortillas to accompany the missionaries' beef and beans.

With the Gold Rush, newcomers flooded the region. Some of the new arrivals saw opportunity in catering to the growing appetites of the miners. Chinese-run eateries predominated in 1849, serving up chops and hashes for the hungry masses. Soon successful miners began to use food to demonstrate their hard-won wealth, and merchants pressed clipper ships into service to supply the burgeoning young city with wheat, pork, preserved fruits, butter, and eggs from Boston or New York; rice, smoked oysters, and tea from China; sugar and pineapples from Hawaii; and alcohol from all over. By the 1870s, the transcontinental railroad provided refrigerated transport between San Francisco and the eastern and midwestern states; refrigerated steamships similarly improved the quality of foods shipped across the Pacific.

In 1857 the Mechanics' Institute began holding regular fairs to promote the city's industrial development. These fairs recognized innovative food technologies, improvements in fruit preservation or viticulture, baking or chocolate manufacturing. San Francisco's Midwinter Fair of 1894 and the Golden Jubilee of 1898 furthered these same goals of promotion and improved production. San Francisco nurtured its image as a cosmopolitan city, offering culinary experiences from all over the world, as well as countless kinds of bread, including tortillas, matzo, milk bread, potato bread, whole wheat bread, graham bread, soda bread, knäckebröd, kulich, Eastern bread, inner twist, family loaf, rye bread, pumpernickel, raisin bread, cornbread,

Figure I.1.   The City of San Francisco: Birds-eye view from the bay looking southwest. Photo Charles R. Parsons, c. 1878. Courtesy of the Library of Congress.

and more than a dozen different French and Italian bread varieties. In 1891 the Women's Educational and Industrial Union gave a great "Feast of Nations," a cosmopolitan fundraiser where amateur cooks shared their culinary traditions with each other. Ladies representing nineteen countries enjoyed Hungarian noodles, Ukrainian dumplings, southern-style headcheese, New England clam chowder, Chinese savory rice, West Indian fritters, Polish stuffed fish, French ragout, Mexican beans, Swedish herring, and many other dishes.[1]

Then the great earthquake and fire of 1906 devastated the city and its ambitious development plans. The disaster also undermined San Francisco's reputation as one of the nation's premier places to visit for self-indulgent pleasures, including (but not limited to) gastronomy. By 1915, however, San Francisco was back on its feet, hosting the Panama-Pacific International Exposition and welcoming visitors from around the country and around the globe for this major world's fair. The 1939 Golden Gate International Exposition showcased the Bay Bridge and Golden Gate Bridge, literally and symbolically connecting San Francisco to the rest of the world. The new bridges prompted people to cross the bay for a good meal, whether that meant coming into the city to enjoy restaurants such as Omar Khayyam's, the Fior d'Italia, or the venerable Jack's; or heading out to Trader Vic's in Oakland, or, later, Berkeley's Chez Panisse. Military personnel coming through the Bay Area during World War II confirmed the region's reputation for celebratory excess and great restaurants, despite the shadow cast by the U.S. government's internment of the city's many Japanese residents. After the war, San Francisco made a priority of attracting visitors back to the city.

From popular columnist Herb Caen to the San Francisco Convention & Visitors Bureau, official and unofficial promoters started selling a simplified story of San Francisco's culinary highlights. Just as the cable car was the city's iconic transportation, sourdough became the city's iconic bread and Crab Louis became the city's iconic salad. People enjoyed their shared rituals: oyster cocktails or Hangtown Fry at Sam's Grill; cioppino at Bernstein's Fish Grotto; sand dabs or rex sole at the Tadich Grill; oysters Kirkpatrick or Green Goddess salad at the Palace Hotel; and Irish coffee at the Buena Vista. Likewise, certain treats created collective memories for a generation or two of San Francisco's children: Blum's coffee crunch cake, Ghirardelli chocolates, It's-It ice cream sandwiches at Playland-at-the-Beach, and hot popovers at the Cliff House.

Only in the 1970s, with the counterculture challenging all these icons, did a renewed California cosmopolitanism emerge, along with the world-renowned California Cuisine movement. Once again, California became known not just for a few tasty treats but for an exuberant approach to food, cooking up a wealth of local produce, dairy, meat, and seafood with well-honed, cross-cultural culinary techniques. It felt exciting and new, even though making the most of Northern California's naturally toothsome ingredients had a long history in San Francisco. As M. F. K. Fisher said, when asked to explain California Cuisine: "It's what we've always had, but we're just more aware of it now."[2]

CHAPTER ONE

~

The Material Resources

San Francisco is famous for its hills and its location on the beautiful San Francisco Bay. But fifteen thousand years ago, the bay was not there—although the hills already were. As the Pleistocene glaciers melted, ocean levels rose and spilled over the foothills of the Coastal Range. The rising waters created an enormous new bay at the mouth of a great river flowing from the Sierra Nevada Mountains and emptying into the Pacific at the Golden Gate. The San Francisco Bay became the largest Pacific estuary in the Western Hemisphere and one of the world's finest harbors. The tip of the resulting peninsula would eventually be recognized as an ideal location for a city looking out to the world.

The region is equally famous for sitting on top of the San Andreas Fault, the Hayward Fault, and other complex faults, where the Pacific Plate and the North American Plate rub against each other. The area from Santa Clara Valley ("The Valley of Heart's Delight") in the south, up through Marin, Napa, Sonoma, and Solano Counties in the north, has been used for agriculture for a short period, compared with most farmland around the world. This soil and the Bay Area's Mediterranean climate produced a rich ecosystem long before humans were planting crops for food.

Offshore, the Pacific teems with fish, shellfish, and marine mammals. Fresh seafood has always played a major role in the city's culinary offerings; the specific dishes have shifted, however, as people overexploited one species after another. Sea creatures enjoy the benefits of the California Current

System, which brings cold, oxygenated water down from Alaska in the summer and brings warm water up from Baja California in the winter helping whales, dolphins, and sea lions migrate north along the coast. The Pacific High, a high-pressure system off the California coast, allows northerly winds in the spring to pull nutrients up toward the ocean's surface to support phytoplankton. In the summer this high-pressure system wards off ocean storms, but in the winter it moves south, so Pacific precipitation comes ashore and waters the land of Northern California. The movement of the Pacific High explains much of the region's seasonal variation between mild, wet winters and warm, dry summers.

Coastal California's hills add even more variation, so people from out of state are often surprised by the contrasts among the Bay Area's microclimates. California's Coastal Range in this area includes the Sonoma Mountains, the Santa Cruz Mountains, the Diablo Range, the Berkeley Hills, the Oakland Hills, and the Marin Hills, whose highest peak, Mount Tamalpais, can be seen from parts of San Francisco, as can Mount Diablo in the East Bay. Within the city itself, the famous hills (Telegraph Hill, Nob Hill, and Russian Hill) and their steep, southern counterparts (Mount Sutro, Twin Peaks, and Mount Davidson) serve a similar role.

These hills channel winds and coastal fogs in regular patterns, generally keeping cool moisture on the western side of the hills and leaving the eastern sides warmer and sunnier. Summer fogs provide impressive microclimates for growing artichokes, brussels sprouts, lettuce, cauliflower, and broccoli, while orchards and vineyards thrive in the hot inland valleys. The hills themselves used to be blanketed with oak trees and evergreens, from giant Sequoia redwoods, to Monterey pines and cypresses, to Pacific madrones (a hardwood with smooth red bark and tasty berries).

At its northeastern edge, the bay becomes a delta where the Sacramento and San Joaquin Rivers empty their waters. Surrounding estuaries and marshes support a wide range of seabirds and shorebirds, fish, amphibians, and small mammals. Before people started filling it in to build valuable real estate, the San Francisco Bay was a third larger in area. The bay supplied Native Americans with many fish and shellfish, as shown by the extensive shell mounds on its shores. After San Franciscans developed a taste for oysters, the bay became home to "seed" oysters imported from Washington State, later from the East Coast, and still later Japan. Jack London got his first taste for adventure as a young oyster pirate, poaching from oyster beds in the bay around 1890. Sadly, rising levels of pollution gradually made the water unsuitable for raising oysters, just as the bay also became inhospitable for shrimp, clams, and many other kinds of seafood.

The Mission Era

Europeans first settled the shores of the San Francisco Bay in 1776, when Hispanic colonists established a military encampment, the Presidio, at the mouth of the bay. Spain, like other European powers, was then expanding its political and economic reach around the world.

In 1542 a Spanish expedition headed by Juan Cabrillo had sailed north past the San Francisco Bay to Point Reyes, and then headed south to Monterey Bay; in 1579 Sir Francis Drake landed near Point Reyes, but, like Cabrillo, the English explorer left without noticing the San Francisco Bay. By the eighteenth century, Spain had taken over much of the Americas and stretched north to control Alta California: today's California, Nevada, Arizona, Utah, western Colorado, and southwestern Wyoming. Over on the East Coast, the Anglo colonies were busy challenging England in 1776. (The new United States did not come west past the Mississippi until after the Louisiana Purchase of 1803.)

In 1774, however, the Spanish explorer Juan Bautista de Anza had forged a land route from the garrison at Tubac (in today's Arizona) to Monterey, California, aided by Native Americans such as Sebastián Tarabal and Salvador Palma. Two years later, Anza's group retraced that route and continued on to San Francisco Bay, bringing seeds, cuttings, and a thousand head of cattle.

In late March 1776 Anza determined the future locations of both the Presidio and the Mission San Francisco de Asís. He noted good potential cropland nearby amid lush, green hills. The group stopped by a lovely creek they named the Arroyo de los Dolores. Chamomile grew on the banks of the creek; fennel and other herbs were nearby, along with fresh springs and plenty of grass for their horses. The explorers sprinkled some seed corn (maize) and chickpea and marked the spot for the future Mission of San Francisco, also known as Mission Dolores. Spanish administrators hoped the region would prove a stable agricultural base, supplying mining districts in the Sonoran Desert and Sierra Madre Mountains. By 1781, however, ongoing conflicts with Native Americans made the overland route impassable for supply trains.

Supply ships only arrived about once a year, bringing rice, oils, sugar, nuts, olives, chocolate, spices, wine, and liquors—but not coffee until around 1830. The ships occasionally supplied other foods, such as crab powder, dried shrimp, lard, and ham. Otherwise, people living in the San Francisco Presidio and the Mission had to meet their own food needs, relying on the Bay Area's rich potential as farmland. As early as 1777, a group of colonists moved down to the South Bay to found the Pueblo of San José de Guadalupe and supply San Francisco with grains, beans, and produce. Settlements along the Pacific coastline and in Santa Clara Valley provided other crops; meat came from *ranchos* (land grants) in the San Bruno Mountains, south of the city.

People living in the region around 1800 shared a basic diet. For the most part, they subsisted on corn, beans (including garbanzos and lentils), chili peppers, and beef. But their different backgrounds led them to different means of varying the routine—people found ways to demonstrate their personal taste and their social class.

The military or religious leaders might start the day with hot chocolate, followed by a hearty breakfast of beef, beans, and corn tortillas or wheat bread. For lunch they would have a good beef broth, thickened with cabbage, rice, or garbanzo beans; soup with vermicelli, tagliatelle, macaroni, or perhaps dumplings; a beef and vegetable stew; and a dessert of cheese, a sweet cake, or other confection. The women would sip a little wine while the men enjoyed a small glass of Spanish *aguardiente*. The evening meal would have beans and chicken or beef again, cooked in a spicy sauce. On Fridays they ate fish: perch, herring, lingcod, and toadfish/midshipmen; all proliferated right offshore from the Presidio.

Ordinary colonists relied more heavily on corn and beans, with beef, rabbit, chicken, and vegetables for flavor. They would start the day with maize porridge (*atole*), or a stew of roasted pumpkin or corn, often with some milk mixed in. Lunch and supper might be some ground corn or wheat, boiled into a gruel with lard, salt, and chili peppers. Occasionally, they might get *champurrado* for breakfast (*atole* made with chocolate), or brighten the day with a bit of cheese—the most popular was *asaderas con panocha* (unmolded fresh cheese with brown sugar).

Most meals were taken in liquid form, as stews, porridges, soups, or gruels, sometimes accompanied by tortillas or, later, by bread. Colonist families only received slightly more food rations than single men, so married women would prepare extra corn tortillas for bachelors in exchange for corn and beans. Native American women were tasked with soaking and boiling corn in lime (nixtamalizing) and then grinding the corn with a roller and stone (*mano* and *metate*)—arduous and time-consuming work. Most of the colonists were of mixed ancestries; very few thought of themselves as pure Indio, African, or Spanish. But they took advantage of being in a new country to emphasize the Spanish parts of their heritage. Whether from lack of familiarity with the region's foods, or to distinguish themselves from the local people, they generally ate no shellfish except mussels, and no wild edible plants aside from hazelnuts, blackberries, and elderberries. They used tableware imported from Mexico, Spain, and China, including water jugs from a particular part of Western Mexico, where the clay was reputed to give water an especially good taste and keep it cold and fresh.[1]

They did rely heavily on the labor of Native American converts, who often arrived at the California missions weak and hungry due to the disruption of their former foodways. Colonial officials banned swidden burns (setting fires to promote new growth); colonists also destroyed wild seed crops by letting livestock graze on the hills, and introduced invasive foreign plants such as wild mustard and wild oats. The missions provided food and integrated new converts into the mission system, but did not explain ahead of time that conversion was irrevocable, or that guards were in place to prevent Native Americans from leaving and returning to their old ways. After epidemics devastated the mission communities around 1800, the colonial military waged aggressive campaigns to capture more local people and rebuild the missions' labor force.

At Bay Area missions, Native Americans breakfasted on a cheaper kind of *atole*, made from roasted barley meal instead of corn, and ladled out into rough bowls made from bark. For lunch, they ate *pozole* (a stew of nixtamalized corn, wheat, peas, and beans, with occasional bits of meat). They received another serving of barley porridge in the evening. A French navy officer reported that the porridge was not even salted for flavor. The crusty scrapings from the bottom of the barley pot went to children who had memorized their catechism well. At Christmas or other special occasions, Native Americans might get sweet potatoes, fried bread, chocolate, and a portion of meat. To roast grain for these various gruels, they used bark baskets, shaking them quickly over coals to get the grains to swell and burst before the basket scorched.[2]

Along with these porridges and occasional treats such as tortillas or fried bread, the baptized local people also incorporated elements from their pre-mission diet. When possible, they went hunting for rabbits and ducks. They gathered shellfish by the shore, they got permission in the fall to go into the hills to collect acorns, and in the spring they toasted all kinds of little edible seeds for *pinole* (coarsely ground meal). Hispanicized colonists often rejected foods they associated with the indigenous people; conversely, the Native Americans made a point of supplementing their diet with a few foods that would remind them of their upbringing.

In 1837 the tiny settlement of Yerba Buena was founded in what is now the North Beach area of San Francisco, as a commercial district by the port. Ten years later, when the United States annexed California, U.S. authorities merged the nearby Presidio and Mission Dolores with the commercial district to form the city of San Francisco.

The Gold Rush Era

Starting in 1848, rumors of rivers and mountains rich with gold drew hundreds of thousands of people to northern California. Miners and other forty-niners ate much better than one would expect, at least when they were feeling flush. Unlike loggers or construction workers (and other largely male labor forces), California's prospectors felt they had left normal life behind in order to achieve not just a solid living but vast wealth beyond their wildest dreams. The quickest way to feel really rich was to splurge on culinary delicacies such as oysters and champagne. Instead of buying fancy clothes or jewelry, they treated their friends to turtle soup and lobster salad.[3]

Not all meals were feasts. Prices for fruits and vegetables soared on occasion. Scurvy was a problem, mostly because people new to the region felt compelled to treat the disease with canned fruit or bottled lime juice, rather than relying on local produce. But local fruits and greens could also be prohibitively expensive: miners on-site sometimes paid a dollar apiece for potatoes and onions and about fifty cents a pickle to supplement a spare diet of beans and bread. Miners paid up to three dollars for a single fresh egg before Petaluma's chicken farms began flourishing in the mid-1850s. High prices called forth ingenious solutions, and usually it was not long before suppliers stepped forward to meet demand at a much lower price.

During the Gold Rush, food prices in San Francisco varied enormously but most ingredients were only sporadically unaffordable. To the extent that miners spent their gold on food, it was more to show off than to avoid pangs of hunger. In addition to dining on champagne, roast curlew, and expensive omelets at the respected Ward House, San Francisco denizens in the mid-nineteenth century also enjoyed excellent French and Chinese restaurants. One miner turned author reported:

> The table d'hôte [set menu] at the best taverns was about three dollars, at others a dollar; at the corners of the plaza, and principal streets, were stalls, where coffee, cakes, pies, etc. were vended to those unable to pay the costs of tavern fare. Some eating-houses resembled our English chop-houses; these were decidedly preferable: each person [ordered] from a printed bill of fare; and if at all voracious, or choice in his selection of food, ten dollars were easily expended . . . Dishes of the most incongruous characters are placed on the table at the same time: boiled and roast meats, fresh and salt, potted meats, curries, stews, fish, rice, cheese, frijolis, and molasses, are served up on small dishes, and ranged indiscriminately on the table; there is a total absence of green vegetables . . . Molasses is a favourite fixing, and eaten with almost every thing. Some of the less refined neither use fork or spoon, the knife serving to convey to the mouth both liquids and solids, which is done with surprising velocity.[4]

BILL OF FARE
WARD HOUSE, RUSSEL & MYERS, PROPRIETORS
SOUTHWEST COR OF CLAY AND KEARNY

THURSDAY, DECEMBER 27, 1849

Soup
Ox Tail .... $ 1 00

Fish
Baked Trout, White and Anchovy Sauce .... $ 1 50

Roast
Beef	$ 1 00	Mutton, stuffed	$ 1 00
Lamb, stuffed	1 00	Pork, Apple Sauce	1 25

Boiled
Leg Mutton, Caper Sauce	$ 1 25	Corned Beef and Cabbage	$ 1 25
	Ham	$ 1 00	

Entrees
Curried Sausages, à mie	$ 1 00	Tenderloin Lamb, Green Peas	$ 1 25
Beef, stewed with Onions	1 25	Venison, Port Wine Sauce	1 50
Stewed Kidney, Sauce de Champagne	1 25		

Extras
Fresh California Eggs, each .... $ 1 00

Game
Curlew, roast or boiled to order .... $ 3 00

Vegetables
Sweet Potatoes, baked	$ 0 50	Irish Potatoes, mashed	$ 0 50
Irish Potatoes, boiled	0 50	Cabbage	0 50
Squash	$ 0 50		

Pastry
Bread Pudding	$ 0 75	Rum Omelette	$ 2 00
Mince Pie	0 75	Jelly Omelette	2 00
Apple Pie	0 75	Cheese	0 50
Brandy Peach	2 00	Stewed Prunes	0 75

Wines
Champagne	$ 5 00	Claret	$ 2 00
Champagne, half bottle	2 00	Champagne Cider	2 00
Pale Sherry	3 00	Porter	2 00
Old Madeira	4 00	Ale	2 00
Old Port, half bottles	1 75	Brandy, per bottle	2 00

 BREAKFAST—*From half-past 7 to 11, A.M.*

 DINNER—*From half-past 1 to 6, P.M.*

 TEA—*From half-past 6 to 12.*

Figure 1.1. Ward House menu, 1849.

He also enjoyed eating in the city's many fine Chinese restaurants:

> The best eating-houses in Francisco are those kept by Celestials, and con-
> ducted Chinese fashion; the dishes are mostly curries, hashes, and fricasees,
> served up in small dishes, and as they were exceedingly palatable, I was not
> curious enough to enquire as to the ingredients.

Norman Assing's Canton Restaurant on Jackson Street was famous in the Chinese community as early as 1849, and by the next year Assing had opened the Macao & Woosung at Kearny and Commercial Streets, serving Chinese food to a broader audience. Similar dining was available at Tsing Tsing Lee's Balcony of Golden Joy and Delight and other Chinese restaurants, marked with yellow silk triangles out front. San Franciscans appreciated their "excellent cookery," all you could eat for a dollar a head. In addition to cheap servings of "chow-chow and curry," these places also offered common American dishes including steaks and pork chops.[5]

A second wave of mining wealth boosted San Francisco's fortunes when the initial Gold Rush was tapering off. Starting in 1859 the Comstock Lode in eastern Nevada produced silver and gold enough to spark new dreams of easy money and the San Francisco high life. Those who found success included Adolph Sutro, who later developed the Cliff House before becoming mayor of San Francisco; William Ralston, founder of the Bank of California; and William Sharon, future U.S. senator from Nevada. Ralston and Sharon built the Palace Hotel (known informally as the Bonanza Inn), though Ralston's fortunes turned and he drowned just before the hotel opened in 1875. In the 1870s and 1880s, cosmopolitan San Franciscans worked to smooth their rough edges, just as their bartenders worked to smooth the harsh taste of the popular pisco brandy. Besides champagne and the new, easy-to-drink Pisco Punch, urban sophisticates also drank Gold Fizzes and Silver Fizzes, celebrating the yields of the Comstock Lode.

Agriculture

As the region's population exploded during the Gold Rush, nearby farms could not keep up with the urban demand, and more and more food arrived from the Central Valley or from beyond California's borders, by ship or railroad. After 1869 the transcontinental railroad connected California to the eastern states. Food moved across the country in just a week, where earlier that trip might have taken six months. Farmers in the Bay Area began growing crops to sell across America and around the world—vegetables such as asparagus, beans, and potatoes; grapes for wine; hops for beer; and orchards full of plums, prunes, cherries, apricots, walnuts, olives, peaches, and pears. At first, stone fruit was sun dried for shipment across the country; after 1870 growers began using sulfur to maintain the fruit's bright colors.[6]

In the 1850s farmers raised wheat in the Santa Clara Valley, but before long wheat took over the Sacramento Valley and then the Central Valley. California wheat was famous around the world for its lightness, texture, and gluten content. Growers produced as much as forty-one million bushels in 1890, just before the devastating economic panic of 1893. Large-scale irrigation, railroad shipping via the Central Pacific Railroad, and technological advances such as steam combines and tractors laid the groundwork for the Central Valley's continued agricultural development. Technology and irrigation could not have achieved such success, however, without cheap farm labor—a continual influx of immigrants willing to perform backbreaking labor under poor conditions for very low wages.

Wine

California's native grapes were too sour for wine, but missionaries in need of sacramental and table wine planted a European variety that became known as the "Mission" grape. Wines from San Jose and Santa Clara were considered decent, but the Sonoma Mission gained a reputation for producing truly excellent vintages. After the Mexican government secularized the mission vineyards in 1830, other property owners in the Sonoma region also turned their attention to viticulture. The rich, loamy soil and mild climate made the North Bay an ideal location for the nascent wine industry in California. And the poverty of many immigrants provided a steady supply of cheap labor for the vineyards: "whole families of them, the babies asleep in empty crates, while men, women, and children filled pile after pile of boxes with great clusters of luscious grapes."[7]

In the 1850s Charles LeFranc modeled his Santa Clara Valley vineyards on those of Bordeaux's Château Margaux, planting Cabernet and Malbec varieties in order to produce fine, blended wines. Around the same time, the Hungarian nobleman Agoston Haraszthy was also planting European varietals to blend with the traditional Mission grapes at Buena Vista, his winery in Sonoma. In 1861 Haraszthy's apprentice, Charles Krug, founded a small vineyard and winery in nearby Napa Valley. By the 1880s, more and more of the region's vintners were turning to the hills, where warmer nights preserved vines from dangerous spring frosts. These mountain vineyards produced more intense wines, which helped make a name for Northern California viticulture.[8]

Before Northern California's wines could find a market outside the state, however, two calamities hit the industry: phylloxera and Prohibition. In the late 1880s vines grown from European stock succumbed to the phylloxera pest, and yields collapsed in California just as they had a decade earlier in France. French scientists discovered that native American grapes may not have been good for wine, but they were resistant to phylloxera. Vintners reconstituted their vineyards by grafting European vines onto native American rootstock. In the late 1890s, University of California viticulturist Arthur Hayne worked with his French counterparts; together they identified a resistant rootstock appropriate for California.

As Californian winemakers rebuilt and excitement about northern California wines began to spread, the second crisis arrived with the passage of the Volstead Act in October 1919. Across the United States, citizens could no longer legally buy wine or other "intoxicating liquors." The wineries in northern California that survived did so by focusing on sacramental wines, raisins, table grapes, or concentrated grape products for home wine making, called "raisin cakes," "wine bricks," or "Bricks of Bacchus." The California wine industry did not fully recover from Prohibition until after World War II, when the University of California at Davis played a large role in creating sanitation guidelines and identifying optimal growing zones for the region's wines. Northern California winemakers began seeking a wider market. Dry wines such as Cabernet Sauvignon and Zinfandel began to push out the sweeter, more forgiving wines. Whites such as Chardonnay and Sauvignon Blanc also got a boost in the 1970s. And when two Napa Valley wines (a Chardonnay from Chateau Montelena and a Stag's Leap Cabernet Sauvignon) charmed the French judges at the famous Paris Wine Tasting of 1976, wine critics and consumers realized that Northern California wines could hold their own at fine meals around the world.

Water

Until the mid-twentieth century, Northern California's agriculture made do with water from rainfall and groundwater rather than irrigation. But San Francisco consumers' water supply was a serious concern for the growing metropolis, leading first to the creation of the Spring Valley Water Company and eventually the Hetch Hetchy reservoir and aqueduct.

In the early days, water wagons delivered barrels of water to San Franciscans, right to one's doorstep—for those who did not live on too steep a hill. The water came from springs in Sausalito and cost about a dollar per barrel. In 1858, after ten years of rapid urban growth, city leaders wanted to expand San Francisco's water system and granted the Spring Valley Water Company a monopoly on the provision of municipal water. The Spring Valley Water Company piped water from creeks and watersheds in nearby San Mateo County, but the cost of water kept rising as the city expanded in the 1870s. In 1878 city residents complained that the company was gouging the public, charging "famine prices" for water and refusing to let private residents dig their own wells to obtain "superior water at less expense." And the next year Bay Area political economist Henry George declared that "water in San Francisco costs more than bread, more than light; it is a very serious item in the living expenses of every family. . . . There is no large city in the civilized world where water costs so much."[9] In 1879 California adopted a new constitution, granting municipalities the right to set water rates. Nevertheless, the Board of Supervisors' new power did not end the controversy over the water monopoly.

More water was needed. San Francisco's industrialists began to look to the Sierra Nevada watershed to meet that demand. By 1899, Mayor James Phelan had hired an insider to evaluate Hetch Hetchy Valley for a dam to supply San Francisco with necessary water. That insider was the engineer Joseph Lippincott, officially tasked with overseeing water resource management in the western states for the federal government—not for San Francisco. In 1910 city voters, disregarding the impassioned pleas of environmentalist John Muir, voted twenty-to-one for a vast aqueduct to bring water from Hetch Hetchy to San Francisco. The program took more than twenty years and about one hundred million dollars to complete.[10] In the meantime, Oakland, Berkeley, Richmond, and other cities on the eastern shore of San Francisco Bay formed the East Bay Municipal Utility District, known as "East Bay Mud," to bring their own water from the Sierras by way of the Mokelumne Aqueduct.

Earthquakes

In recorded history, three great quakes have struck the Bay Area. The first was an 1868 quake on the Hayward Fault with an estimated magnitude of 7.0. Mark Twain reported that the streets filled with people in the blink of an eye. He spread the rumor that many of them were pouring out of saloons, even though it was not even eight in the morning. A saloon keeper on Clay Street apparently called out to people in the street, "Come in, gentlemen, and take something, for the next shake brings her down"; and, indeed, newspapers mentioned the "large quantity of wines and liquors" destroyed by the shaking. The joke of the week went something like this:

— Say, Bill, did you hear of the great shake this noon?
. . ???
— Why, over at the corner saloon. A party went in there and shook five sixes for the drinks.

The city was self-consciously amused by its reputation as a hot spot of drinking, gambling, and earthquakes.[11]

That quake was known as the "Great San Francisco Earthquake" until April 1906, when the city was hit by a 7.9 earthquake on the San Andreas Fault, followed by devastating fires. Many small fires started amid the shoddy wooden construction south of Market Street; with water mains broken by the earthquake, the fires spread uncontrollably. North of Market and west of Van Ness, the "Ham and Eggs Fire" started when a family attempted to cook breakfast without realizing the damage the earthquake had done to their chimney. The grand restaurant Delmonico's originally survived the earthquake, but soldiers taking a break from rescue work accidentally started a fire that spread to the nearby Alcatraz and Orpheum Theaters and then joined other fires into a huge conflagration.[12]

Food supplies posed a serious problem immediately after the 1906 earthquake. Mayor Eugene Schmitz ordered grocers, dairymen, and bakers to provide supplies for soldiers to distribute to the earthquake's survivors, many of whom were slowly trying to make their way out of the city. Sugar, coffee, bread, canned goods, and prepared meats were distributed out of Golden Gate Park. Mothers without food offered beer instead to their hungry children, but vast stores of hard liquor were seized by the police and poured into the gutters as a preventive measure. The Presidio sheltered more than thirty thousand people, handing out flour, beans, and canned goods, along with milk and eggs for families with young children. In other parts of the city, wagons from the Spring Valley Water Company circulated twice a day because many of the city's pipelines and mains were broken. Milk wagons sold their milk for only five cents a quart, and food sellers in general followed advice to keep their prices low. By the fourth day after the disaster, refugees were cooking meals over little campfires in Golden Gate Park. In succeeding weeks, Washington State sent seventy thousand pounds of fresh beef; Oklahoma City sent two boxcars full of flour; the Iowa Canning Company sent twenty-four thousand cans of corn; and cities all over the country took up collections for San Francisco.[13]

San Francisco's Chinese cooks ended up taking care of their stranded employers:

> There sat four people on the ground eating fried pork, potatoes, and Chinese cakes. . . . [They said,] "Charlie has been with us many years. . . . Yesterday we were without food, and Charlie disappeared. . . .Toward dark he came back with a bamboo pole over his shoulder and a Chinese market gardener's basket suspended from either end. . . . [He brought] an assortment of pork, flour, Chinese cakes and vegetables, besides a half-dozen chickens and a couple of bagfuls of rice. Charlie had been foraging in Chinatown."[14]

A telegram from the U.S. Department of War stressed that the Chinese community should receive the same care and resources as other residents of San Francisco. This federal concern came at a time when city officials were in fact evaluating the opportunity to move Chinatown out of the heart of the

Figure 1.2. Street kitchens jokingly named for famous hotels, 1906. Courtesy of the Library of Congress.

city. Within six days of the Great Fire, a committee was in place to consider moving Chinatown to Hunters Point in order to seize that prime downtown real estate. The Chinese government used diplomatic pressure to end this relocation scheme.[15]

By June, two of the most famous restaurants had reopened: Delmonico's and Tait's. Rumors spread that others would be opening soon, but some of the rumors merely reflected humorous signs that people posted over their improvised outdoor cooking encampments—"The Poodle Dog," "The Pup," "Zinkand's," "The Palace Grill," and "The New St. Francis." Three months after the earthquake, the real St. Francis was hosting banquets again, and the Buon Gusto Italian Restaurant offered "Ravioli Every Day." In November, the Old Poodle Dog Hotel and Restaurant opened on Eddy Street, near Van Ness. Slowly, the social life of San Franciscans started to get back to normal.[16]

The next major earthquake came in 1989, on the San Andreas Fault. This quake killed sixty-three people, took down several freeways, and caused immense property damage across the Bay Area, although not nearly as much as the great earthquake and fire of 1906. The San Francisco Conservation Corps provided meals to displaced earthquake victims in the Tenderloin. The California Grape and Tree Fruit League delivered free fruit to Glide Memorial Church. People came together to help each other, occasionally harking back to the amiable 1868 earthquake spirit, as when security guards watching over the city's shops let hungry San Franciscans take six-packs of beer as well as food.[17] A Brazilian poet then living in the Bay Area wrote about the connection between food and comfort, which emerged from the earthquake a bit shaken up:

> The house stood as a metaphor
> for love and food, and when it shook,
> out gusted a perfume of Ovaltine
> and holiday turkeys and bacon rinds,
> and the vomit of sick children on the stair.[18]

As San Franciscans look ahead to the inevitable next big earthquake, they know they might not have access to food, water, and electricity for days or even weeks. Disaster professionals recommend everyone keep on hand sufficient supplies to survive a devastating earthquake without state or federal help for at least three days. The hope is that when the Big One comes, the damage will be limited, leaving people able to rebuild quickly and to enjoy the small pleasures that come from sharing food and drink with fellow citizens.

CHAPTER TWO

~

Native American Foodways

Coyote gave the people the carrying net. He gave them bow and arrows to kill rabbits. He said: "You will have acorn mush for your food. You will gather acorns and you will have acorn bread to eat. Go down to the ocean and gather seaweed that you may eat it with your acorn mush and acorn bread. Gather it when the tide is low, and kill rabbits, and at low tide pick abalones and mussels to eat. When you can find nothing else, gather buckeyes for food. If the acorns are bitter, wash them out; and gather 'wild oat' seeds for pinole. . . . Even though it rains a long time people will not die of hunger."

Coyote killed salmon and put them into the ashes to roast. He did not want his children to eat them. . . . Once in a while he reached into the ashes, took a piece, and ate it. Then his children cried out that he was eating fire and would be burned. When they wanted to take some, he did not let them. He said "You will be burned."

Jacinta Gonzalez (b. 1838) and Maria Viviana Soto (b. 1823), Monterey, California[1]

These stories, collected at the turn of the twentieth century, reveal clues to Native American foodways in the Bay Area. Nuts and seeds make up more than half of the specific foods mentioned above and indeed nuts and seeds were a staple of the Native American food supply, as far back as the oldest archeological evidence. The mention of wild oats serves as a reminder that those foodways changed over time, as the Spanish brought oats to Northern California in the eighteenth century. Over centuries, local people developed new technologies to process nuts and seeds and make them into ever tastier dishes, accompanied not just by seaweed but by other delights of sea and land, shifting both by season and by epoch.

Archeological Evidence

The earliest indications we have of humans in California suggest they migrated across the Bering Strait from northeastern Asia around fifteen thousand years ago and then some of them headed south along the North American coastline. Around the same time, the San Francisco Bay Area as we know it was coming into existence, with the ocean rising to flood the grassy valley that now lies at the bottom of the bay. This peaceful area offered sustenance to the foragers who traveled the marshes around its edge, finding frogs, shorebirds, and mollusks to eat. The foragers also wandered up into the hills to hunt elk, deer, and smaller animals. They ate seeds from needle grasses and may have sweetened their food with milkweed sap.

Around ten thousand years ago, people in the Bay Area began using milling slabs and hand stones to process small hard seeds and nuts; they also shaped rocks into projectile points for hunting and made hooks for fishing. Over the next few thousand years, they developed the mortar and pestle technology, which let them turn nuts and seeds into flour. Over the same period, they refined their basketry skills, from simple containers woven from sea grass to watertight baskets made of sedge root and willow shoots. Three thousand years ago, they also figured out how to leech acorns to eliminate tannic acid. And they weighed down their nets with grooved-stone net sinkers to catch Pacific herring and other fish.[2]

By two thousand years ago, Bay Area Native Americans had replaced their nets and net sinkers with barbless fish spears—suggesting that they had overfished for Pacific herring and similar small fish. This was a golden age of coiled baskets and shell mounds, as people began to settle in semisedentary communities. They had time to weave careful baskets and to eat piles and piles of shellfish. Over centuries, Native American shellfish consumption in the Bay Area seems to have shifted from predominantly oysters, to mussels, to clams, to tiny horn snails, perhaps because again they overharvested each kind and had to move on from a favorite food source to one they could still find in large quantities. These semisedentary groups developed regional migration patterns, hunting for young deer in the upland meadows in summertime, harvesting acorns in the fall, and hunting sea otters, harbor seals, and the flightless scoter (a kind of goose-sized duck with oversized, appetizing "drumsticks") along the coast during winter months.[3]

The Early Historical Record

"Ohlone" is an outsider's term for indigenous people living around the Bay Area, much like the Spanish term *Costeños* or the anglicized version, Costanoans—all different ways for others to refer to these coastal people. About forty different groups were living in the Bay Area, speaking eight to twelve different languages. By the time outsiders started writing about these people, however, the Native American experiences had changed enormously, and many lived in or around the various missions, particularly in San Rafael, Santa Clara, San Juan Bautista, Santa Cruz, and Carmel. Remaining records across this broad region can help illuminate Ohlone foodways and culture.

The Ohlone and other Northern California people had a food system based on technologies unfamiliar to the Europeans who first encountered them. Pyrodiversity was of these technologies: Native Americans burned grasslands in a loose rotation, every five or ten years, to keep the brush from growing thick and eliminating their favorite seed-gathering spots. Fires had other positive effects: they provided nutrients to the vegetation; attracted animals to young plants; and helped with the regrowth and maintenance of a variety of plants, such as manzanita, which the Ohlone appreciated for its berries; and gray pines, a good source of pine nuts. When the Ohlone set the fires themselves, they helped prevent the unpredictable devastation of huge forest fires.[4]

Basketry was another such food-related technology. The Ohlone used baskets of many kinds to gather, preserve, and cook their food. Baskets served as all manner of pot, container, and tool for these non-metalworking people. In 1579 Sir Francis Drake landed near Point Reyes in Marin County, and described the local baskets as "so cunningly handled, that the most part of them would hold water."[5] When the Spanish arrived, they noted numerous kinds of watertight baskets among the Ohlone: some like barrels for storing water; others like canteens for carrying water on a hike; others like pots for heating soups and porridges, and small watertight dippers for scooping out the soup once it was cooked. The soup baskets could not go over a fire, but Ohlone cooks placed hot rocks into the soup "pot"; to avoid burning the basket they kept the rocks constantly moving, swapping in red hot rocks every so often.

The Ohlone used curved basketry paddles to beat ripe grass seeds into burden baskets for transport; and stored seeds, nuts, and roots in lidded baskets at home. They used basket traps to catch fish, birds, and small animals. They wove several kinds of basket trays, some for winnowing seeds, some for toasting seeds (by tossing them around in the tray with hot rocks), and other trays for serving food. They even wove special bottomless baskets that fit onto a stone mortar base, preventing bits of food from scattering when pulverized with a pestle.

Nuts and Seeds

Native American techniques for turning nuts and seeds into a food staple made a big impression on outside observers. Europeans who arrived in the late eighteenth century noted the porridge- and bread-making methods of the Ohlone and other nearby peoples. These people dried acorns in the sun until the bitter kernel could be popped out. Women sat on flat bedrock worn down by years of grinding, or they placed a heavy stone mortar between their legs. With piles of kernels at their side, they raised the heavy stone pestle up and down, up and down, crushing the kernels to the beat of simple songs such as

Bellota—a—a, bellota; mucha semilla—a—a, mucha semilla.[6] [Acorns, acorns; many seeds, many seeds]

That verse was already reworked into Spanish from the original, either because the mission father who jotted it down asked for a translation or because the women themselves were comfortable using Spanish in daily life. Possibly Native American women living at the mission took particular pleasure from singing in Spanish about crushing acorns as they brought down their heavy stones in time to the song.

After grinding the kernels into a fine meal, women scraped the meal into a shallow sifting basket, shaking the basket to separate out any coarser particles. Once all the acorn meal was well ground, the women dug a small pit in the sand and lined it with leaves or pine needles. They leached out the acorn's bitter taste by spreading the meal in that depression, then pouring water over it and letting the water seep through the meal over hours, adding more water from time to time. Once the bitter taste was gone, the women still had much work to do. Bread makers scooped the wet meal from the sand and placed it in a deep, watertight basket. They added a succession of hot rocks to cook the meal—"it bubbles and pops and boils"—until it turned into a thick pudding.[7] The pudding was scooped into small molds and plunged into cold water to set into the consistency of stiff blancmange. The Ohlone ate that soft bread with their fingers, often in combination with seaweed or clams, or else they fried the bread until it was hard. They also relished "acorn chips," crispy little bits that stuck to the hot rocks after they were retrieved from the bubbling pudding.[8]

In the 1770s Francisco Palóu, the first pastor of Mission Dolores, remarked that Native American women living in the region worked hard gathering grass seeds in season. They ground the seeds with acorns or hazelnuts, or sometimes with buckeye seeds if the acorn crop had failed, turning the mixture into flour for mush or a doughy bread. (Native Americans leached out buckeyes' natural poisons by baking or boiling the seeds for hours.) He also noted a black seed (probably chia seeds, a source of protein and omega-3 fatty acids, or else red maid seeds, which are also oily); the Ohlone ground these black seeds separately and made them into fat tamales, the size of an orange—"These are very savory and taste like toasted almonds, but are very greasy."[9]

In 1786 Jean-François de Galaup, comte de Lapérouse, sailed into Monterey Bay and found a warm reception at the Spanish mission in Carmel. He watched Ohlone women at work in the mission and reported that they had switched from acorn and grass seeds to corn and barley, but still spent most of every day in front of a grindstone:

> They have no means other than using a cylindrical stone to crush the kernels on top of another stone. M. [Paul Antoine Fleuriot] de Langle, who witnessed this process, gave his own mill to the Spanish priests; he could hardly have been more helpful: four women now can do the work of a hundred, and they will have time left over to spin wool from the mission's flock of sheep.[10]

But when another French officer arrived in Monterey ten years later, he was stunned to see Native Americans still squatting in front of a broad, flat stone, crushing kernels with a large prism-shaped stone. He wondered why they did not use a European-style mill to do the job. The local governor explained that de Langle and de Lapérouse had left a simple mill for them to use as a model for building more: "and yet, despite my encouragement and my orders," said the governor, "no workers wanted to put their hands to the job."[11] It was an odd turn of phrase. Did the governor mean that the local Native Americans did not want to use the mill? Or that mission craftsmen did not want to build new mills? Or they were not capable of copying the European technology? In any event, the Ohlone women continued to make flour with the help of a grindstone, just as they had long before Europeans arrived.

The Ohlone who were not living in the Spanish missions continued to use many kinds of wild seeds and herbs in their *pinole*—peppergrass seeds, tarweed seeds, shepherd's purse, red maids, clarkia, sunflowers, as well as the new import, wild oats. They boiled *pinole* with water into a mush or porridge, or mixed it with acorns to make an even more substantial porridge or bread. The Ohlone toasted other seeds in basketry trays and ate them separately from the *pinole*—not just black chia seeds but also California laurel seeds and pine nuts.[12]

Seasonal Migration

Like earlier people in the Bay Area, the Ohlone moved around with the seasons, following the food supply. Ohlone people worked in groups to harvest large amounts of food in a short period of time. They shared some of the food immediately in celebration of the harvest, where it also served to demonstrate the power and prestige of prominent families. The annual acorn festival was always a big event, and no one wanted to miss the feast that followed the spawning of the salmon. In addition to devoting themselves to regular bursts of work followed by feasting, Ohlone also prepared much of the seasonal bounty to be stored until winter.

Even though they moved through the seasons to gather available foods, they did not take the same paths to the same places every year. The acorn crop in one area might be much worse than last year, whereas the next hill over had a bumper crop; sometimes people would spend the acorn-gathering season in the tanoak groves, seeking the tastiest nuts, even if the valley oak acorns were more plentiful. The live oak acorns were good during hard times, because they were very filling, but on the other hand they were very time consuming to leach, and not nearly as good as black oak acorns. In some years, the acorns never really came in, and the Ohlone ate leached buckeye seeds instead. The best place to fish could also shift significantly from year to year, and even favorite berry patches and grasslands did better some years than others, depending on large-scale weather patterns such as El Niño/La Niña and local microclimates. Rather than commit to a precise trajectory, the Ohlone paid attention to how young plants were developing in various prospective harvest areas each year. Scouts traveled around to check the productivity of different nut groves, grasslands, and hunting and fishing spots, and then planned where and when to return with a larger group of harvesters.[13]

In early spring, the Ohlone gathered young greens, such as clover, poppy, tansy-mustard, melic grass, miners' lettuce, mule ear shoots, cow parsnip shoots, and the very young leaves of alumroot, columbine, milkweed, and larkspur. Some were eaten as salad greens, some were steamed or boiled. In the spring, women also used sticks to pry up many young bulbs, such as Mariposa lily, California harebell, soaproot, and brodiaea (Indian hyacinth), eating the bulbs either browned over hot coals or roasted. They also dug up wood fern rhizomes and cattail roots. Later in the spring, Ohlone women walked through the Bay Area's grasslands, sweeping seeds into the burden baskets on their backs: tansy-mustard, sage, chia, evening primrose, clarkia, madia, and red maid. At the end of the day, they might throw a handful into the air, saying to any helpful spirits, "I send you this so that another year you will give me greater abundance."[14]

In summer, the Ohlone wandered Bay Area hills harvesting fruits and berries, such as wild strawberries, wild grapes, wild cherries, elderberries, huckleberries, madrone berries, manzanita berries, thimbleberries, and toyon berries. The Ohlone ate these fruits fresh off the plant, but also cooked them into jam, dried them for winter, or made them into nonalcoholic ciders. They ground up cherry pits and used them for flavoring other dishes. Spring and summer were good times to watch for salmon spawning in Bay Area rivers.[15]

Summer was also a good time for hunting black-tailed deer, antelope, and elk in the forests. Ohlone men wore elaborate deer heads with antlers, using the disguise to help them approach deer close enough to make the kill. During hot summers, animals came down into the valleys after the higher regions became brown and dry. Hunters roasted a downed elk right on the spot, allowing them to carry it home in pieces. They usually roasted deer back in the village, or dried it for storage until winter. Like most Northern California Native Americans, they divided a deer up according to set guidelines—the hunter gave the deer's liver to an elderly woman who helped raise him; the skin to his wife; the sirloin and legs went to his family, his wife's family, and their neighbors; and the stomach was made into a blood sausage, stuffed with other organs, blood, and salt and shared among the other hunters. The bones, head, and shoulders went to those with less of a claim—but they were welcome gifts, as everyone enjoyed the nutritious and delicious bone marrow.[16]

In late summer Ohlone people often hunted grasshoppers, setting fires to herd large numbers into pits for roasting: "They cannot jump out so easily as they jumped in."[17] The Ohlone ate them mashed into a paste, ground up to add to the day's acorn mush, or roasted like shrimp; alternatively, they dried the roasted grasshoppers for the winter.

In the fall, the hunts continued, but everyone took several weeks off for the great acorn harvest. Once scouts had reported on the best locations, men, women, and children traveled together to set up camp in the forest. Children climbed the trees and shook the branches until acorns rained down on their elders. Meanwhile men used sticks to knock the nuts down. Then everyone stooped to pick up the acorns, snap off their caps, and toss the nuts into burden baskets. They brought the acorns back to the village and enjoyed a great feast. Most of the acorns were saved for later in large woven granaries suspended on stilts to keep squirrels away. During good years, they tried to gather enough for two years' worth of acorn bread, in case the next acorn harvest was disappointing. The fall was also a time for digging up wild onion tubers and camas bulbs. Ethnographer Mabel Miller tried camas and liked it, describing the bulb as "about the size of the little finger, shaped like a sweet potato, and with much of the same flavor." The Ohlone pit roasted wild onions and camas bulbs to flavor their acorn bread or mush.[18]

During the rainy winter, the Ohlone people mostly stayed close to their village, gathering mushrooms and enjoying the shellfish bounty they found in and around the bay, such as bay mussels, oysters, and clams. The Ohlone cooked black turban snails in their shells in hot ashes, then smashed them open and shook them in a tray to separate the meat from the shell. Whether littlenecks or long necks, bent-nose or Pismo, clams were also cooked in hot ashes or surrounded by hot rocks until their shells opened up. Limpets were cooked by shaking them in baskets with hot rocks. The Ohlone considered mussels dangerous in the summer, but the rest of the year they enjoyed them boiled, steamed, or roasted. Once the shell opened, the meat could be eaten right away or dried for storage.

Migrating ducks and geese were frequent visitors during the winter months and provided many a feast, along with game birds such as pigeons and quail. When Pedro Fages explored the San Francisco region in 1770, he encountered a group of friendly Ohlone who accepted his glass beads and gave him decoy geese stuffed with straw, "which they employed to catch an infinite number of these birds."[19] The Ohlone also sought out seabirds such as cormorants, pelicans, and gannets—in Spanish such diving birds have been called "alcatraces," and the name Alcatraz derives from their prevalence on local islands. To the locals, gulls tasted bad, so those birds were a food of last resort. The Ohlone relied on the bay and the local rivers and creeks for many kinds of fish, from Pacific herring, sardines, pilchard, and minnows (including the hitch and the thicktail chub which is now extinct) to larger fish like the Sacramento splittail, the Sacramento sucker, trout, salmon, and sturgeon. Depending on the fish, the Ohlone might use basketry traps, nets, seines, weirs, harpoons, or clubs. Sometimes they stunned fish with the juice from the roots of soap plants, and collected them when they floated up to the surface. Many of the fish were broiled or dried, whereas their roe were boiled instead.[20]

As winter turned back to spring, the marshes around the bay expanded with the melting of the winter snowpack. In those moist areas Ohlone hunted beavers, river otters, and small animals such as raccoons, gophers, squirrels, and rabbits, along with shorebirds such as herons, curlews, sandpipers, and dowitchers. Turtles and their eggs were always popular. Ohlone people sometimes cooked them in their shells, placing the turtle upside down in a shell-sized little pit and covering it with hot ashes. When it was done, they cracked open the shell to eat the turtle meat.[21]

Coastal Visits

Ohlone used seaweed and other marine plants for salt and as part of their diet. Some preferred the leafy parts of various sea plants, while others preferred to bake the stalks in hot ashes, or sun dry cakes of seaweed to fry and eat later. Some seaweeds were dried and then pounded and eaten with acorn mush.

On seasonal trips to today's Ocean Beach or elsewhere along the coast, they would pry abalone off rocks and scoop it out with a hard wooden spatula, to be eaten fresh or dried. Greg Sarris, whose ancestry includes Coast Miwok and Kashaya Pomo, said his family relied on abalone harvest songs to keep them safe from the ocean:

> These songs and how to do them were the property of certain families. . . . You had to know the abalone songs and perform them correctly, otherwise . . . you'd be swept away by the tide, and brought to the underwater world where you would have to live like the abalones.[22]

They speared octopuses and crabs, or poked them with a stick and waited for them to grab on. They ate the tentacles of the octopus, roasted over a fire; they baked crabs in hot sand or roasted them before eating both the crab flesh and its roe. They also ate other invertebrates, such as gooseneck barnacles, sea anemones, and sea urchin roe. They waded into the surf with spears to hunt shark, rays, and ocean salmon. They used nets or their fingers to catch rockfish, cabezon, surf smelt, and surfperches. Smelt were washed in salt water, dried in the sun, and then smoked to preserve them for winter. Surfperches, however, were usually cooked right away in hot ashes or earth ovens and eaten right there on the beach. The Ohlone also sometimes hunted large marine mammals, such as harbor seals, sea lions, and sea otters, and a whale that beached itself provided a welcome opportunity for a feast.[23]

After the Missions

The mission period was very disruptive to Native Americans living in the San Francisco Bay Area. By 1782 two Franciscan fathers at Mission Santa Clara reported:

> Animals, both large and small, belonging to the townsfolk have caused unceasing damage to the crops put in by the Indians. . . . They will have to rely for

their food on the herbs and acorns they pick in the woods—just as they used to do before we came. This source of food supply, we might add, is now scarcer than it used to be, owing to the cattle; and many a time the pagans living in the direction of the pueblo have complained to us.[24]

Ohlone people needed time to harvest a bounty of food whenever and wherever it came into season. They also needed access to grasslands and forests full of seeds and nuts, as well as waterways teeming with fish, shellfish, and shorebirds. Spanish cropland and pastureland interfered with those resources, and the later American presence was even worse. Historians estimate there were eleven thousand Ohlone before the mission period; by 1832 fewer than two thousand remained, and the numbers continued to decline until the twentieth century. The first generation of Mission Indians had taken occasional vacations back to their old villages during peak harvest seasons and maintained some ties to their families and traditions. Their children growing up in the missions never learned the old foodways. They became dependent on wheat and barley, and, later, potatoes and coffee, in place of acorn bread, *pinole*, and manzanita cider. Many Ohlone died of disease at the missions; of the ones who lived through the secularization of the missions in the 1830s, many either married outsiders, found sporadic work on local ranchos, or otherwise eked out a living in isolation from their original culture. They could not go back to their ancestral villages, as those villages no longer existed.[25]

Over the decades, the Franciscans had to go further afield to recruit labor to the missions. They brought in people from outside the Bay Area, from tribes such as the Miwok, Yokut, and others. These new arrivals adjusted less well to mission life and escaped more often. Perhaps they knew that their old foodways were still practiced back in their villages in the Central Valley and the foothills of the Sierra Nevada mountains. After secularization, these people faced new problems that came with the American presence in Northern California: mining operations clogging the rivers, farmers draining the marshes, ranchers fencing grasslands for pasture, and American hogs eating all the acorns. Americans also readily adopted the mission-era forced labor techniques, and many large ranchos looked like feudal estates with peonage in place.[26]

General Mariano Guadalupe Vallejo was the largest landowner (*ranchero*) in the North Bay after 1830 and prided himself on treating his Native American workers well. He made sure his workers had thick *atole* to drink morning and night before they went to work his fields, and he assumed their dependents (children and the elderly) would somehow figure out how to obtain wild game, nuts, seeds, roots, and berries to diversify the family's diet. When George Simpson visited in 1842, he was not impressed:

> We visited a village of General Vallejo's Indians, about three hundred in number, who were the most miserable of the race that I ever saw. . . . Every face bears the impress of poverty and wretchedness. . . . They are badly clothed, badly lodged and badly fed. . . . As to food, they eat the worst bullock's worst joints, with bread of acorns and chestnuts, which are most laboriously and carefully prepared by pounding and rinsing and grinding. Though not so recognized by the law, yet they are thralls in all but the name.[27]

Another observer complained about similar abusive labor practices and paltry nourishment on Northern California ranchos:

> The process is, to raise a posse and drive in as many of the untamed natives as are requisite, and to compel them to assist in working the land. A pittance of food, boiled wheat or something of the kind, is fed to them in troughs, and this is the only compensation which is allowed for their services.[28]

By 1864, when the *Daily Alta California* announced a grand Indian festival to be held in Calistoga Springs, the newspaper predicted this would be the last one in California: "This Indian gathering will afford a fine opportunity—probably the last—of witnessing the aborigines of California, who are fast disappearing and dying away."[29] The Native Americans who remained kept a low profile.

When Europeans began settling by the San Francisco Bay in the late eighteenth century, the hillsides were covered in a tasty herb, the "yerba buena" for which San Francisco was named until the Americans seized control and changed the name in 1847. The Ohlone taught Europeans to make a minty tea from this herb. The Bay Area was then, as now, a land rich with resources for eating well. Sadly, interactions among people in the Bay Area have often looked more like the rancheros giving hungry Native American workers a trough full of boiled wheat, and less like people from different cultures sitting down to share a friendly cup of herbal tea.

Twentieth Century

Many people view Native American culture in the twentieth century as predominantly reservation based, if not virtually extinct. And yet a vibrant intertribal American Indian culture emerged in the Bay Area in the mid-twentieth century. Some of the first Native Americans to arrive in the Bay

Area from outside California came from the Laguna Pueblo reservation in New Mexico. Men came first, to provide labor for the Santa Fe Railroad Company during a strike action in 1922; their families followed soon thereafter and stayed for decades. In Richmond, they lived in boxcars on sidings; and yet the residents managed to grow corn, melons, and other familiar fruits and vegetables right in the rail yard. Residents built outdoor communal ovens to bake Indian breads, mostly from wheat or corn rather than acorns or buckeye seeds. The men went deer hunting together and roasted venison for the community. During World War II, Native Americans in the armed services would come by the "Santa Fe Indian Village" with others from their units: "Our wives . . . fix the food, Indian food, for these service boys."[30] The Native American poet Wendy Rose, who was born in Oakland, wrote about Hopi lamb stew cooking in the Richmond Indian Village, uniting the people with its luscious smell even as the steam fogged up the windows of the boxcars.[31]

A more overtly political approach emerged in the 1960s as a response to the federal government's official relocation program. As Adam Fortunate Eagle wrote:

> We formed the United Bay Area Council of American Indians, and set a charter for ourselves almost as a new and unique tribe the government had never anticipated. . . . We met at the Friendship House every Wednesday night at 7:30, or as near to that hour as the invariably late "Indian time" would allow . . . Sometimes [this was] like asking the lions to talk it over with the lambs. Some tribal differences ran deep. . . . It was the actions of the government trying to eliminate us that served to bring us together.[32]

Wednesday night potluck dinners at Oakland's Intertribal Friendship House became famous across the country. People from all tribes were welcome at this meal, which often featured chili beans, beef stew, and fry bread: "Everybody wanted fry bread, at least once a month or so. So they would have the fry bread at the Friendship House, and that's when the women who know how to make it would make it."[33] Sarah Poncho was one of those women. In 2000 she sat down with Susan Lobo and explained her technique, adapted below.

Wednesday night dinners at the Intertribal Friendship House were the best-known gathering, but on other nights people would get together at Café La Boheme in the Mission or the Hilltop Tavern on MacArthur Boulevard in Oakland. The Hilltop was apparently where activists met to plan the American Indian occupation of Alcatraz in 1969. Once they had taken over the island on November 20, they began immediately planning for a major Thanksgiving feast the next week.

> [We were] planning what would be the most memorable Thanksgiving in San Francisco history, and maybe the West Coast's answer at last to the Massachusetts affair that started the whole thing. The United Council and the Ladies Club of the Friendship House were quickly galvanized to the idea. Indian women and couples volunteered to do the cooking, others offered help in assembling and organizing the feast. . . . Bratskellar Restaurant in fashionable Ghirardelli Square announced they would furnish and prepare all that was needed for a meal of major holiday proportions. . . . They were going to feed the Indians on the occasion celebrated for the time when Indians fed the pilgrims. . . .
>
> Many carried dishes and pots of their own to that walled pit of an exercise yard behind the cell block: fry bread and mutton, venison and corn, food for the feast that came with personal attention and meaning beyond the meal itself. By the time the boat from the Bratskellar arrived, brimming with roasted turkeys and trimmings of the finest imagination, as many as four hundred Indians had gathered on the island.
>
> [It] seemed to me the biggest and best family gathering of all time. The drums began to sound. . . . The rhythm and the singing reverberated off the concrete walls, bouncing back and swirling again so you could feel the music in your chest, almost prodding you, demanding that you get up and dance.[34]

This Thanksgiving feast hooked the national media, and brought a spotlight to Native American hopes and demands both in the Bay Area and around the country. In the end, the Alcatraz Occupation lasted almost nineteen months, ending in June 1971. And Oakland's Intertribal Friendship House is still going strong, serving as a center of oral history, and simultaneously building a future based in large part on community gardening, community dinners, and community cooking classes.

Sarah Poncho's Fry Bread: Mix four cups of flour with two tablespoons of baking powder and a teaspoon of salt. Pour half a cup of warm milk into a cup measure, fill with warm water, and stir. Slowly add the liquid to the flour mixture and mix until the dough comes together. Then form it into small balls and roll them out. Heat a quarter inch of vegetable oil in a heavy frying pan, and fry each round: "When it is bubbly and golden color, turn it over. Then take it out and let the oil drip off; put it on a paper towel and eat it!"[35]

CHAPTER THREE

~

Immigrants and Ethnic Neighborhoods

Immigrants have played an enormous role in San Francisco's history, and their different cuisines are a large part of the city's charm. The Gold Rush sparked a wave of immigrants from around the world, each bringing memories of their favorite foods. Over decades, immigrants integrated their cuisines into the city's colorful culture, while more recent arrivals continued to bring new tastes from afar.

Californios/Latinos

After the Native Americans, the next earliest migrants to the San Francisco Bay Area were the people who came north from Mexico during the time of the missions. After a generation or two, these Indio, African, and Hispanic colonists began calling themselves Californios. But the community remained in flux, as the Gold Rush brought a flood of new immigrants from Mexico, Central America, and South America. Non-Latino San Franciscans referred to all of these people, whether long-established in California or new immigrants, as "Mexicans," particularly when they saw them lounging on the Plaza (Portsmouth Square) in serapes, drinking in a saloon, or speaking Spanish together. Members of this community, while recognizing their diverse backgrounds, also acknowledged their cultural commonalities, often referring to each other as fellow Latinos.

San Franciscans were at first respectful toward Californio rancho owners, especially because these rancheros offered many travelers a night's hospitality and a hot meal. Spanish missions had helped launch the ranchos, lending them cattle and Native American labor. Rancheros focused on raising livestock, although they also cultivated a little corn, wheat, barley, beans, olives, melons, onions, and peppers. In the early days, condiments and groceries were hard to come by. Californios enjoyed hearty but simple meals, cooked over hot coals or in small adobe ovens. At daybreak, ranchos offered the household and any guests *desayuno*: a mash of corn, milk, and sugar. By the 1830s, many were waking instead to a small breakfast of coffee and buttered biscuits. Mid-morning, the household would enjoy some roast beef and beans (*frijoles*), along with tea or perhaps a second cup of coffee. At noon, the cook would put out a large spread, beginning with a beef or mutton broth; then Spanish-style thick soups with rice, noodles, or dumplings; then a spicy stew or *puchero*, made with meat cooked earlier in the broth along with a simple sauce of dried peppers in the winter, or a more elaborate sauce in the summer composed of minced peppers, tomatoes, onions, parsley, and garlic. They might also enjoy enchiladas or a *pozole* made with nixtamalized corn, pumpkin, peppers, and pigs' feet. Those who were still hungry could sate their appetite with corn tortillas and *frijoles*. The tortillas were thin cakes of meal, sometimes of wheat, but usually of corn, patted between the hands and cooked before the fire or on thin sheets of iron. The *frijoles* were boiled and then usually fried with lard, serving as a cheap staple throughout the day. To combat the heat in the afternoon many people drank a cup of tea, although men often preferred a stiff drink of *aguardiente*. When trade picked up with the Gold Rush, little extras such as sugar and chocolate were easier to get— so, naturally, teatime became more elaborate, with cakes, cookies, jams, curds, and biscuits on the table. Late at night, the rancho household might gather again for a light supper of beef ragout and more beans. The cutlery and dishes were a bit haphazard, given the frontier lifestyle, and horn forks and spoons were common, along with whatever knife a person was already carrying for utility.[1]

Special occasions called for more elaborate dishes, such as bull's head baked with the hair on. One wealthy miner raved about the taste:

> The finest thing that mortal ever put between his teeth is a cow's head baked with the hair on. Its sweetness can only be dimly guessed at by people who have never eaten it. First, our Mexican cook—it's a Mexican dish, I believe, or a dish of Mexican origin . . . trims the head of the horns and other portions that he may see fit. Then he incases it in clay so that it is air tight, and buries

it in a bed of red hot coals. I don't know by what process nature produces her results, but she brings forth a dish in this rude way that all the Frenchmen who have ever been born can't touch. The head is allowed to bake until it is perfectly done. Then it is carefully removed from the fire, the clay knocked off and the most confirmed dyspeptic can make such a meal as would astonish his friends. You can't help eating. The juice fairly pours out of the meat and the flavor is something exquisite.[2]

The miner brushed aside the cook's hard work with an offhand statement about "nature" producing these exquisite results, besting the achievements of the finest French chef. But the intense, moist flavor of the dish comes through in his description.

Elk was another favorite treat, especially for those who had the thrill of hunting it. In August 1893 a journalist for the *San Francisco Chronicle* went along on one such hunt with a local ranchero, whom he referred to as "Mexican, or 'Californian' as he is locally styled." The elk came down to the valleys in search of food after a hot summer, making them easy prey:

The Mexican elk hunter is "armed" only with a lariat and a luna, a crescent-shaped knife which is tied at the end of a slender pole about ten feet in length. . . . The Mexican sits calmly on the horse, talking sarcastically to the struggling elk, bestowing praise upon himself and smoking his cigarette. After the animal exhausts himself, the Mexican throws the luna and hamstrings the elk.[3]

The hunters would bring the elk back to the rancho to feast on the white meat or sell it to "Mexican" restaurants in San Francisco, of which there were quite a few.

In the early 1850s, customers could already enjoy "fiery dishes dear to Mexican palates" at Pascual Estrada's Fonda Mejicana, on Jackson between Montgomery and Sansome.[4] Another Mexican Fonda was located on Pacific near Dupont Street (now Grant Avenue). In the early years, those restaurants were mostly frequented by Californios and by the many new Latino immigrants from Mexico, Peru, and Chile, who lingered over their meals, enjoying the Spanish in the air and scooping up *frijoles* with a bit of tortilla. In 1862 Hipólito Jurado ran the Restaurante Mejicano on Dupont, where he served "everything that may be Mexican-style food," according to his advertisements. His successor, Antonio Domínguez, renamed it the Restaurante del Aguila de Oro (Golden Eagle Restaurant) and reached out to non-Latinos as well, advertising that one could get not only Mexican hot chocolate but also tea, coffee, and cake. By September 1863 the business had changed hands again, and was now owned and run by a Chilean immigrant named Estéfana Abello de Stocking (also known as Fanny Stocking). Other Mexican restaurants also tried to reach a broader clientele, offering customers an option of bread instead of tortillas.[5]

Non-Latino San Franciscans took note of this appealing cuisine. The city became known for its tamale vendors in the 1880s and 1890s, and people of every background looked forward to their appearance in the late evenings. Steam rising from their pails, vendors would call out "Tamales, tamales—fresh tamales!" Many people had a street vendor to thank for providing their first taste:

> The tamale is eaten in a small way by persons of all classes in this city, though in many instances the first eating has been purely experimental and with doubts as to results, and a certain recklessness is often necessary before one can make up one's mind to such a repast. It is with surprise therefore that many persons on eating a well-cooked specimen have remarked how their peculiar taste has pleased them, and these have voted without hesitation that a good tamale is a good thing and not to be despised as an article of diet.[6]

Once they got a taste for tamales, people began seeing them as more than just street food. By 1888 tamale makers were selling most of their stock to restaurants and saloons. Rumors about small-scale street vendors using seagull in their "hot chicken tamales!" may have given people a reason to seek assurances of higher quality. The *Chronicle* published numerous articles playing up the "mystery" inside the tamale's corn husk, and other stories explaining the ingredients used by the best tamale manufactories: fresh chicken boiled with cumin, oregano, and salt; cornmeal mixed with leaf-lard and butter; spicy *chile colorado* (red chili sauce); and a couple of pickled olives. Robert H. Putnam, founder of the California Chicken Tamale Company, sent his vendors out in hygienic white uniforms, rather than in ethnic garb. In 1895 when the corn harvest failed and a pound of corn husks rose from two cents to forty cents, many producers went out of business, whereas others were inspired to start canning tamales instead.[7]

Quite a few new restaurants opened to serve fresh tamales and other Mexican cuisine. Besides tamales, diners at these establishments enjoyed what a reporter called "tortillos and frijoles," a dish he described for non-Latino readers as "a good deal like baked beans and hot cakes," except that "the beans are cooked in a style impossible of imitation by the Americans, and the cakes are as sweet as a nut, and contain more of the natural grain than an American stomach is used to."[8] With the new clientele, restaurants began accommodating American tastes and expectations, and food writers

Figure 3.1. Fresh, hot, chicken tamales! From *Street Types of Great American Cities* (Chicago: Werner Co., 1896). Courtesy of the Department of Special Collections and University Archives, Stanford University Libraries.

began distinguishing between, on the one hand, the delightful dinners one could have at fun places such as Luna's Mexican Restaurant on the corner of Dupont and Vallejo, and, on the other, the "real thing," served at various Mexican stands on Broadway.[9]

The colorful writer Yda Hillis Addis, who spent her childhood traveling through Mexico with her itinerant photographer father, made an effort to explain Mexican cuisine to non-Latino San Franciscans:

> To my mind no system of cooking, the French not excepted, equals the achievements of a first-class Mexican *cocinera*. . . . I defy anyone not to like the genial, flat, Naples-yellow wheaten tortillas, some half-inch thick and six inches in diameter. . . . Those tortillas still held the sunlight and the fragrance of the wheat fields. And, moreover, they were not only toothsome, but nutritious, since the entire grain of the cereal had been ground up for the material of their making. The famous enchiladas are simply fresh tortillas, fried in lard or oil, with a liberal allowance of green or red chile peppers, reduced to a sort of gravy, after which they are sprinkled with grated cheese, finely chopped onions, and sometimes other herbs, with an olive or two, and then they are neatly rolled ready for serving. Tamales are made of bits of meat, of beef, pork, chickens, or what not, also dressed with chile and perhaps olives, and covered, dumpling fashion, in a lump of maza or corn dough, the whole wrapped in a neat oblong parcel in corn husks, tied with a bit of the same, and boiled like dumplings.
>
> In the cooking of beans the Mexicans, as a rule, have no superiors, if they have equals. . . . They are so savory and strengthening that one can well understand how so large a proportion of the population find in them their principal aliment. The variety used is a rich brown bean . . . and the best of all is the Vera Cruz sort, absolutely black in color.
>
> After washing the beans are put in an *olla* or earthen pot (and the use of clay instead of metal vessels is one of the advantages of the Mexican method). . . . It is best to cook them a long while on a very slow fire. . . . When boiled sufficiently the beans are ladled into a pipkin in which lard is boiling and in which have been thrown a few shreds of onions, and they are stirred and crushed with a wooden spoon until they are as stiff as jam.[10]

Helpfully, she included a line drawing showing readers how to use fried tortilla chips to scoop up their *frijoles* in a Mexican restaurant.

In San Francisco, tourists and locals began to visit "Little Mexico" on the southern part of Telegraph Hill.[11] They were amazed that one could see tortillas made "daily, nightly and hourly," with the same repetitive gestures, grinding the nixtamalized corn into *masa*, patting it into rounds, which were then "flung on a thin sheet of iron heated by charcoal and as quickly flung off." Visiting this poor Mexican neighborhood in 1896, journalist Lillian Purdy noted that "some of the women . . . are neat about themselves and their work, and you would not hesitate to eat one of their tortillas." She was impressed with these industrious women, working from three in the morning until late at night, selling tortillas for a penny apiece to the local Mexican stores. The store owners sold tortillas, tamales, beef jerky, *chorizo*, and imported dried fruits such as plums and *guamuchiles*, along with other groceries to nearby restaurant and to households around the block. They also sold imported Mexican pottery, which was said to give the food a special flavor—particularly the prized bean pots and water jugs. Purdy was perplexed but agreeable when a grocer offered her some fruit:

> Have you ever had a Mexican request you to sample his fruits? He will not give you the fruit in your hand, but, asking you to raise your veil, he will hold up the fruit before your eyes and then daintily place it in your mouth, without, of course, touching his fingers to your lips. . . . You must accept his attention graciously.

This little fruit-tasting game explored dangerous questions of race, gender, and class, but with subtlety, to avoid making the players feel uncomfortable—or so Purdy suggested. Some other visitors to this part of town did feel uncomfortable. One complained that the tortillas tasted like "a section of a tramp's shoe fried in soapsuds." But even he admitted that this was not true of the tortillas eaten by the more "aristocratic" section of San Francisco's Latino population, the ones who lived outside Little Mexico.

Over time, the Latino community's center of gravity shifted south to the Mission district. From 1910 to 1930 many immigrants arrived from Central America, pulled by their connections with the coffee trade and other food industries. Mexicans also came in large numbers, escaping the Revolution of 1910–1920. The Mission district offered Latinos affordable housing in addition to bilingual restaurants and grocery stores. During the 1930s, however, the community shrank due to the Great Depression and the U.S. government's Mexican Repatriation Program—an effort to free up jobs for non-Latino Americans. Starting in 1942, the U.S. government bracero program brought many Mexicans back to California as contract labor. Latino immigrants began moving with their families into the Mission district, until by mid-century the neighborhood was about half Spanish speaking. Mission residents could choose among numerous *tortillerías* and *panaderías* and *mercados* offering plantains, yucca, mangoes, sweet limes, *nopales* (prickly pear pads), baked goods, Mexican cheeses and sausages, and "a rainbow of dried beans, herbs and chiles."[12]

Great ethnic restaurants were also a major part of the Mission district's food scene. In 1974 the neighborhood heard rumors that a McDonald's was coming to the Mission. Two local women painted a huge mural, "Para El Mercado," on the side of Paco's Tacos. The mural represented Latinos enjoying food in many ways and was intended to inspire passersby to use their dollars to support local ethnic cuisine. The appeal worked. Among the local landmarks are Roosevelt's Tamale Parlor; Casa Sanchez (which started as a tortilla factory before becoming a taqueria); La Palma Mexicatessen (known for grinding nixtamalized corn into fresh *masa*); La Victoria Panadería; the Nicaraguan bar El Tico Nica; La Taquería (on Mission), El Toro Taquería, and two restaurants with competing claims to have invented the enormous "Mission Burrito"—La Cumbre Taqueria and El Faro. With hyperbole as outsized as the burritos themselves, one *New York Times* writer reveled in his new discovery:

> A whole dinner could be stuffed inside one steam-heated, bargain-priced white flour wrapping. The enlargement of the burrito to humongous, Americanized proportions may be the Mission's supreme contribution to Western civilization.[13]

African Americans

A black man and a black woman were among the first entrepreneurs in San Francisco, where both made money by serving food. William Alexander Leidesdorff, a successful merchant captain, came to Yerba Buena from the West Indies in 1841. Five years later, he established the city's first significant public hotel: the City Hotel on the corner of Clay and Kearny. His hotel served basic fare: plain boiled beef, dumplings, "indifferent bread, and worse coffee." Until his death in 1848, Leidesdorff was a prominent local figure, and a millionaire to boot.[14]

Mary Ellen Pleasant arrived in the early years of the Gold Rush, bringing significant funds and a reputation as a great caterer. She took a position as cook and housekeeper for the employees of Case, Heiser & Co., a commission merchant house on Sansome. A charismatic woman, she made connections with the traders and brokers whose meals she catered, and over the next few decades she built a sizable fortune. At different times, she owned several boardinghouses as well as laundries and other businesses. Pleasant hired a black staff to serve her boarders fine wines and hearty meals. She often did the marketing herself, a thin woman in a plain black dress with a clean white kerchief over her shoulders and a straw bonnet tied under her chin. Vendors at the Washington and Sutter Street markets considered her exceptionally shrewd at haggling down the cost of her provisions.[15] Stories of her southern-style dishes referred to a "Hoppin' John" made with black beans, easier to source in San Francisco than the traditional black-eyed peas, and a popular molasses sponge cake with a tart lemon filling. Pleasant's 1901 autobiography

Figure 3.2. Mary Ellen Pleasant. Courtesy of the San Francisco History Center, San Francisco Public Library.

noted laconically that her mother was from Louisiana but did not mention whether they cooked together.[16]

By 1854, more than one thousand people of African descent lived in San Francisco. They came from northern states and southern states, from Mexico, Peru, Chile, and the West Indies. The city had at least two black-owned restaurants, twelve bars, and two billiard saloons serving drinks and light snacks. In addition, about two hundred black people made a living as cooks in various restaurants, clubs, or private homes.[17] Decades later, stories circulated about the Gold Rush days, when a black woman named Charlott Callander fed sailors at her boardinghouse, and a black Jamaican ran a beloved eatery on the waterfront:

> Jim built himself a shack over the tide-water at a spot that is now the corner of Folsom and Beale streets, four blocks from the water front. . . . All Jim had to do to secure shell fish was to walk out fifty yards or so at low tide, with a bucket, and scoop up all the clams or cockles he desired. . . . It was the most wonderful dinner of its place, price, and period, and it reflects much upon the epicurean perspicacity of the pioneers to learn that the worthy purveyor had to turn away scores of customers daily.[18]

In the 1850s, most African Americans lived on the eastern slope of Telegraph Hill, in cramped quarters alongside Mexicans, Chileans, and other poor populations. In contrast, middle-class African Americans lived scattered throughout the city, usually near their place of business.

Although California joined the union in 1850 as a free state, harsh laws prohibited blacks from voting, testifying in court, or helping fugitive slaves come to California. During the Civil War, President Lincoln relied heavily on gold from the western states. From 1861 to 1865, the federal mint in San Francisco shipped about $30 million in gold east each year. After the war, freed slaves with valuable skills left the South, some heading to California. Abby Fisher made her way from Alabama to San Francisco, where she became famous for her cooking and her pickling skills, honed on a southern plantation.

In 1880 Fisher won two medals at the San Francisco Mechanics' Institute Fair, for best pickles and sauces and best assortment of jellies and preserves. Her pride in her reputation shines through her 1881 cookbook, *What Mrs. Fisher Knows About Old Southern Cooking*. The title alluded to the "Mammy" stereotype of the instinctive plantation cook, but in the book Fisher noted the awards she had won in state culinary competitions. Her professional accomplishments helped Fisher undermine the assumption that black women cooked only by instinct rather than by training.[19]

In the decade after the Civil War, African Americans became established in San Francisco's service industries. When the Palace Hotel opened in 1875, the manager hired 150 African American waiters. The prestigious hotel featured a French head cook, a chief confectioner from Milan, a chief baker from Vienna, and an African American baker from the East Coast: "an old Negro the fame of whose egg-muffins and corn-bread has made him the aristocrat of his race for the last half-century."[20] Less than a decade later, however, those opportunities were vanishing. With the rise of the White Cooks and Waiters Union of the Pacific Coast in the 1880s, many of the service jobs in San Francisco's hotels, restaurants, bars, and clubs were no longer open to African Americans. By 1888 even the Palace Hotel no longer employed black cooks or servers, and an editorial in the *Daily Alta California* newspaper declared: "The object of this movement is to do away with colored help altogether and to have only white men in the kitchen and dining room."[21]

The coming decades were difficult for African Americans who lived in San Francisco proper:

> There are but few colored people in business of any kind today in the City of San Francisco. . . . With but few exceptions, all the avenues of trade are closed to the Negro workman through the powerful influences of the trade unions who rule San Francisco.[22]

After the 1906 earthquake, many blacks moved to the East Bay in search of affordable housing and economic opportunities. In Oakland, more African Americans owned their own businesses or worked in the professions than in San Francisco. By 1929, Oakland was home to more than one hundred black-owned businesses, including fifteen eating establishments. A strong culture emerged of middle-class values and community building. Various organizations offered classes where young black women could obtain culinary training, either to get a job or to feed their own families healthy meals. And during hard times the Linden Street YWCA in West Oakland (known as the "Colored Y") offered food for the hungry as well as referrals for those out of work.[23]

In 1910 a number of East Bay black women joined with women from Southern California to put out the *Federation Cook Book: A Collection of Tested Recipes Contributed by the Colored Women of the State of California*. The only San Franciscan to contribute was Isabelle Barraud, a native of St. Louis and a member of the local Baptist church. She sent in twelve recipes, including squash soup, corn chowder, oyster fricassee, green pepper and grapefruit salad, and stuffed dates. Three black women from Berkeley submitted recipes, ranging from salmon croquettes to jambalaya and one called "Creole Dish":

> **Creole Dish**: Boiled macaroni or spaghetti. Set aside. Take meat (any kind) left over from day before, grind, fry with chopped onions, parsley, celery; add tomatoes, season to suit taste and mix in macaroni.

Eight Oakland women each submitted a recipe. The most prominent partici-pant was probably Hilda Tilghman, a second-generation Californian whose mother, Hettie, ran many of the African American women's social clubs in the Bay Area. Hilda herself was the first black salesclerk hired by Safeway in Oakland.[24] Her contribution to the cookbook collection was a veal dish:

> **Veal Entree (Original)**: Made from the meat left over from the soup pot. Select a firm piece of meat for your soup. When about cooked tender remove from the sauce pan. Cool thoroughly. With a sharp knife cut into small squares. Prepare a small piece of onion and some celery; chop together very fine. Beat 1 egg light and prepare enough cracker crumbs for the amount of meat; pepper and salt according to the taste. When all has been arranged dip your squares into the egg, then into the mixture of onion and celery and last into the cracker crumbs. On the stove have a frying pan containing either hot butter or lard; into this place your veal squares allowing them to brown well and serve.

A woman named Leonya Jones sent three recipes: peach salad, "Strawberry Fluff," and a creamy marshmallow dessert called the "Berkeley Special." Eth-yle Jones sent in a recipe for asparagus:

> **Asparagus on Toast**: Drain the cooked asparagus and cut off the tips laying these on strips of buttered toast. Garnish with slices of hard-boiled egg and serve with cream sauce.

The cookbook did not explain how to cook the asparagus before assembling it with toast and slices of hard-boiled egg. As with most community cookbooks, these recipes assumed a shared familiarity with common cooking techniques.

Hilda Tilghman's grandmother, Lucinda Tilghman, had lived at 662 Fifth Street in West Oakland. After her husband died in the late 1870s, Lucinda took in boarders including a Pullman porter and his son. (Oakland's loca-tion at the end of the transcontinental line meant that railroad workers and porters played a major role in the local community.) Archeologists have uncovered a treasure trove of artifacts from an 1880 rubbish heap in the privy complex behind the house.

Dining was formal, as was the tea and liquor service. Meals featured high-priced beef loin steaks and roasts, ham, and leg of mutton. Mrs. Tilghman also served cheaper meals of ribs, pork shoulder, soups, stews, spareribs, and pig's feet. Bones of elk, goat, ground squirrel, and rabbits were also found in their privy, along with those of chicken, duck, northern pintail, and grouse; some of these may have been acquired on hunting trips.[25]

Lucinda Tilghman also served expensive store-bought fish, including sardines, white bass, Chinook salmon, and California barracuda. Her boarders may have fished for jacksmelt and rockfish, remains of which were also found in the deposit.

Other Pullman porters lived in a boardinghouse closer to the rail yard. Food bones found in a disused well represented almost two thousand pounds of meat, mostly from cheaper cuts. The American Railway Union strike in 1894 did not prevent African American porters from maintaining a high quality of life:

> They ate a varied diet of fresh foods intensively prepared by the cook . . . Meals were served in a formal, fashionable setting that included expensive pieces for use when entertaining with tea and alcohol. Fresh flowers and bric-a-brac graced the parlor. . . . These porters translated the elegance of the Pullman cars . . . into [their] domestic surroundings . . . perhaps as symbols of civility and personal dignity.[26]

Two decades later, another Pullman porter and his wife lived two doors down from Lucinda Tilghman's residence on Fifth Street. They also took in boarders, although household meals were less formal than in the Tilghman boardinghouse. Still, they ate well and used a range of china sets and festive glassware for serving tea and alcohol to their guests.

As African Americans escaped the Jim Crow era of the South and sought out economic opportunity, many of them came to California. A vast migration of blacks from Arkansas, Mississippi, Louisiana, Texas, and Oklahoma changed the Bay Area's culinary culture forever. Migrant women found that in addition to working full-time, they faced other challenges, such as cooking in substandard kitchens and sourcing ingredients to make familiar dishes. Early-twentieth-century migrants arrived with a taste for jazz, and nightclubs sprang up to allow people of all backgrounds to appreciate this new genre—clubs that catered to a mixed audience were called "Black and Tan." Until the 1920s, most Bay Area jazz was performed in the Barbary Coast neighborhood of San Francisco, but Oakland also had a vibrant nightclub scene as early as 1918.[27]

In 1940 industrialist Henry Kaiser established four major industrial ship-yards on the northern tip of the East Bay. North Richmond came to life. Jimmy McCracklin's sister-in-law, Willie Mae "Granny" Johnson, opened the Savoy Club, serving southern-style ribs, chicken, and greens, along with beer and wine. The food was laid out informally on wooden tables covered with oilcloth and went well with the down-home music. Starting after the war, Minnie Lue's, another North Richmond nightclub and restaurant, be-gan offering customers a taste of southern "home cooking" along with blues music. Owner Minnie Lue Nichols liked her job:

> It was just easier to take care of a club . . . than it was to work all day and night in the [ship] yards, then turn around and work at home taking care of babies. Besides, if women was going to cook, they could make some real money off it.[28]

Minnie Lue's club was popular for decades. But the most vibrant night spot in the area was Tappers Inn, owned by George Bally, an East Indian immigrant. Bally served mostly southern food, despite his South Asian background, and maintained a lavish setting with a dance floor, fine white tablecloths, and shiny red leather banquettes. In addition to these visible nightclubs, some African Americans operated secret "after-hours" clubs in their homes. Pa-trons enjoyed great music and real home-cooked meals along with less savory pleasures such as gambling and prostitution.[29]

Across the Bay Area, this new population of southern blacks encountered an atmosphere of distrust and fear. The prewar community of African Ameri-cans had found ways to maneuver around discrimination by white-run hotels, restaurants, and other public facilities. The new arrivals, unused to local ways, faced harsh criticism from the white press and other public voices. Bars and hotels put up signs stating, "We Refuse Service to Negroes." An editor of the *Oakland Observer* complained: "[W]e see, in Oakland . . . white women waitresses serving Negroes in white men's restaurants." Black Californians also resented the southern migrants for bringing in new tastes and smells, new music, and loud religious practices: "Most migrants . . . didn't accept us, and we didn't accept them." During the war years, migrants were crowded together, with sometimes eight families sharing a kitchen and bathroom. They planted familiar vegetables in any available space and shared produce along with marketing suggestions. Willa Henry remembered:

> [I was] so green and upset trying to shop at first. I found some of the things I was accustomed to, but back home we had fresh pork and chicken and greens and black-eyed peas. I went to the Tenth Street Market and asked for some okra and they looked at me as if I was nuts.

The Housewives Market in Oakland was a welcome exception, an early retailer of Creole ingredients such as cornmeal, okra, collard greens, black-eyed peas, and *filé* (a seasoning made from dried sassafras). After annual visits to the South, migrants brought back smoked sausages, hams, or jars of preserves—a taste of home to savor the rest of the year in California. Twin sisters Cornelia Duvernay and Carmelia Chauvin remembered looking forward to care packages from the South: "Our brother in New Orleans would send us crates of shrimp and crab. We had guys from the base who were from Louisiana come over. Mama would cook, and we'd have beer and dance." Migrants mostly ignored the ministrations of middle-class women, sustaining themselves with a culture of informal hospitality, community, and mutual obligation.[30]

San Francisco had many of the same obstacles and opportunities as Oakland. Dr. Wesley Johnson Sr. came from Texas; he opened the Texas Playhouse nightclub in 1942 and helped bring the "Juneteenth" celebration, commemorating the Emancipation Proclamation, to San Francisco in the 1950s. His son vividly remembers an incident in the 1930s, shortly after they had arrived in San Francisco, when the family was excluded from a restaurant on the corner of Geary and Powell: "I didn't understand what was going on and asked my mom 'Why can't we eat here?' She was crying. 'The man says he's not going to serve us.'"[31]

During World War II, the government forcibly relocated the Japanese residents of the Western Addition, sending them to internment camps. Their sudden removal opened up opportunities for new African American migrants to move to that neighborhood. The poet Maya Angelou has written about this period: "Where the odors of tempura, raw fish, and cha had dominated, the aroma of chitlings, greens and ham hocks now prevailed . . . The Japanese area became San Francisco's Harlem in a matter of months."[32] Black businesses began to open alongside Chinese restaurants, particularly in the Fillmore district. Former mayor Willie Brown spoke about his youth, spent in the Fillmore:

> There were many black barber shops. There were barbecue pits all over the place. . . . The bars were basically owned and operated by black people. You had the Texas Playhouse, the Blue Mirror, the Booker T. Washington Hotel, the Virginia and Kansas City Hickory Pits, the Big Glass, Jimbo's Joint. You had places where black people gathered no matter where they resided in San Francisco.

People went to hear hot new jazz musicians, but they also compared notes on the food—discussing whether the sweet potato pie was better at Jimbo's Bop

City or at the Kansas City Hickory Pit, and whether barbecued ribs should be made Texas- or Louisiana-style. Despite these arguments, residents agreed that food and music provided opportunities to build a new community:

> We had to educate the merchants in the Fillmore District because we spoke another language. We had to translate: dry salt bacon meant salt pork in San Francisco. Black eyed peas were beans at that time. The two most important types of fish were catfish and perch; shrimp was not one of our regular feasts. And then we had pork, chitlins, pig's feet, and the jowl, which was the head of the hog. We taught the merchants about things that we were commonly cooking.[33]

African Americans came to San Francisco with a culinary background reflecting their origins. Restaurants and nightclubs in the neighborhood offered customers smoked meats, fried chicken, slow-cooked greens, and cornbreads—both fried and baked. By the 1960s, the Western Addition was almost three-quarters black, and the food culture reflected the diversity of black tastes. One could go to Virginia's on Divisadero for its version of mustard and collard greens, chitterlings, short ribs of beef, ham hocks, sweet potato pies, and deep-dish peach cobblers. Or one could try newer flavors at Behind The Scene (at Clay and Fillmore), which offered vegetarian soul food dishes, such as "soul pie," a soybean and sweet potato pie.[34]

French

When word spread in 1848 about gold deposits in the Sierra Nevada Mountains, French men were among the first to arrive. French traders and adventurers had previously traveled from Louisiana to Sonora in northern Mexico, and from there they were well placed to respond to the stories of easy gold. The ones who found restaurant work more profitable than mining established themselves along Commercial Street, Clay Street, and Dupont Street, in what became known as the Frenchtown district of San Francisco.[35]

Two French friends, Auguste Morinoux and André Benadou, apparently opened San Francisco's very first French restaurant in May 1849. André did the marketing and cooking while Auguste managed the business, located on Kearny Street. Feeling as close as brothers, and both hailing from the south of France, they named their restaurant Les Frères Provençaux. Léon Dingeon was another early French immigrant, offering "delicacies of the season" at his restaurants: Barnum, then Martin's, then the Maison Dorée on Kearny. The Maison Dorée's chef, Flavian Berton, was also from the south of France.[36]

These restaurants served southern specialties such as bouillabaisse made with San Francisco's excellent fresh seafood. Another Gold Rush–era restaurateur, François Mondelet, entertained his patrons "in European style," first at the French House Restaurant and then at the restaurant Lafayette, while his wife, Marie Duchene, ran the Café du Commerce. Duchene described her establishment as "similar to the coffee saloons in Paris," in that she would serve ice creams, broths, rice and milk, and rice soup every evening.[37]

By 1854 numerous French restaurants dotted the city, three on Dupont Street alone. These early French restaurants tended to take evocative names, such as Napoleon, Richelieu, A la Croix Rouge, A l'Esperance, Au Coq Hardi, Au Gamin de Paris, and Au Rendezvous des Ouvriers. Some simply went by French Restaurant or Restaurant de France. One of the most respected French restaurants in San Francisco was called the Jackson House. The famous Poodle Dog started as the Union Rotisserie and Restaurant before acquiring its unusual name. Another early French restaurant was called the Louisiana Rotisserie, founded by French people whose southern flair hailed from the Gulf Coast rather than the Mediterranean.

Isidore Boudin started the first French bakery in San Francisco during these hectic years when the miners were happy to find a good French loaf and escape the routine of biscuits, flapjacks, and "hard bread which we eat half-cooked."[38] Boudin Bakery is still a San Francisco landmark. In addition to popularizing French bread among San Franciscans, Boudin and other French bakers played a crucial role within the French colony, supplying not only the familiar French loaves for daily meals but also traditional cakes such as the *galette des rois*, eaten during the season of Epiphany. In the late 1850s, four bakeries called themselves simply "French Bakery." Frenchtown also had a New Orleans Bakery and a Louisiana Bakery, a reminder of the diversity of San Francisco's French community.[39]

In the small French colony, the earlier arrivals provided assistance to succeeding waves of newcomers, often helping them get low-end jobs in the wine shops, groceries, bakeries, butcher shops, tripe sellers, dairies, boardinghouses, and restaurants owned by other French immigrants. The majority came from the south of France or from the provinces of Alsace and Lorraine. Some went into agriculture, growing grapes in the Santa Clara Valley or Napa Valley, or planting urban gardens with radishes, lettuce, endive, cabbage, turnips, fava beans, artichokes, and peas for local grocers and restaurateurs. The French community continued to gather and speak French (or French dialects) in their clubs and private gatherings and over simple family meals such as onion soup, salad, and grilled chicken. But they were eager to make friends, business partners, and customers outside their own networks.

They adapted to what they saw as odd American customs, such as not letting customers sit down in cafés and bars:

> The remarkable thing about these establishments is that there are absolutely
> no chairs or benches. . . . Americans don't sit down at tables as we French do;
> they are served at the counter and they leave right afterward—to tell the truth,
> they are back before long.[40]

French café owners picked up right away on the idea that a quick turnover
was good for business, especially if customers came back again soon for more.
French store owners faced a constant need to balance appealing to a mass
market, the lowest common denominator of American tastes, with marketing French products and cuisine as luxury goods. Generally, French shops
emphasized champagne and expensive canned goods such as caviar and lobster to bring customers in the door, while also making sure to stock popular
items, such as jars of cornichons mixed with tiny pickled onions and pieces of
cauliflower. The French community encouraged the idea that San Francisco
was the "Paris of the Pacific"—a pleasingly cosmopolitan conceit.[41]

California's French residents took pride in their cooking. However, those
who had left hotel living behind often employed servants in their home
kitchens, resulting in cross-cultural tensions. One French traveler reported
on a visit with his friend Pierre, who had lived in San Francisco many years.[42]
Socializing in the kitchen, the guest asked the Chinese cook his age and got
an insolent reply—"A hundred." Pierre then dragged the cook into another
room and proceeded to give him a thrashing. The traveler followed up by
noting that the cook had been quite docile ever since ("supple as a glove"),
and expressing how lucky Californians were to have Chinese people who
would work for one-fourth the cost of a white person. The racial assump-
tions of the day meant some people faced routine violence and low pay as a
matter of course.

French San Franciscans did sometimes acknowledge Chinese rivals in the
restaurant business. As one French booster put it in 1854:

> The Chinese showed up here with their pots and pans and opened the first
> restaurants. But it didn't take us long to overshadow these preparers of birds'
> nests. The French culinary arts are appreciated all over the world and even the
> coarsest taste buds, far off on the Pacific Coast, still favor us and enjoy our ways
> of preparing all sorts of foods—even with ingredients unfamiliar in France, the
> land of Vatel and Carême. In the beginning, French canned goods were dis-
> pensed prolifically as if they were essential to survival. . . . French restaurants

still serve canned truffled partridge, canned mock turtle soup, and canned peas, but now they serve them alongside other [local] dishes.[43]

The majority of French restaurants in nineteenth-century San Francisco served regular prix-fixe menus with little choice. For two dollars in 1869, one could get soup, a fish course, some simple appetizers (maybe pickled beets, stuffed mushrooms, or a bean and tomato salad), a delicate pâté, then the main course (often a roast), and finally dessert or fruit, and coffee, with a good wine to accompany the meal. One restaurant reviewer wrote that French restaurants are known for their "high prices; dishes with foreign names attached; and waiters who pronounce the said foreign names with decided aplomb."[44] Another commentator joked about the grandiose names of French dishes, imagining a *côtelette d'iguanodon* (iguanodon cutlet), a *"tarte à la couronne bouleversée"* (overturned crown tart), a *consommée de reynard* (fox soup), and a variation on a Charlotte Russe he termed a *"Charlatan Russe."*[45]

Even though French restaurants were at the forefront of San Francisco's culinary scene, people still liked to make fun of them for being a little hoity-toity. Restaurants took that into consideration. Advertisements tried to reassure the public, mentioning fine wines and oysters, as well as each restaurant's liberal hours and reasonable prices. Promotional materials did not mention dishes by their French names, except when advertising in French-language publications. In the restaurant itself, where the French-speaking and English-speaking communities came together to dine, the staff worked hard to make everyone feel comfortable regardless of their background.

Nevertheless, restaurant critics wrote newspaper pieces describing French restaurants as two-faced, presenting an immaculately clean front, with pristine white tablecloths and a roast turning on a gleaming spit and giving off an appetizing smell. In the back, said the critics, the real kitchens were not so glamorous or clean:

[We see] clean, sleek cooks, in white caps, fronting full on the street, with the evident intention of deluding mortals outside into the belief that . . . a black, hot, frizzeling, frying, odor-ladened region does not exist in the rear, where the actual cooking is done.[46]

This was part of an ongoing concern that elaborate French presentations acted to obscure the poor quality of the ingredients used.

Similar insinuations were made about Chinese restaurants. One French writer joked about dishes he supposedly saw on the menu at a Chinese restaurant in San Francisco: cat cutlet, dog soup, roast dog, dog pâté, braised rats.

Americans themselves saw both Chinese and French dishes as potentially deceptive: hiding poor-quality foods with seemingly sophisticated sauces.[47] It was easy to mistrust any substance covering up or disguising the basic food on their plate. Nevertheless, many San Franciscans did appreciate the skill that went into creating these delicious and varied dishes, both French and Chinese.

Chinese

Before the Gold Rush, Chinese men had visited San Francisco's harbor as ship cooks and traders, but the first Chinese immigrant was probably Maria Seise. Seise was originally from Guangzhou (Canton), then lived in Macao, Hawaii, and then Hong Kong, where she began working for the wife of an American trader named Charles Gillespie. Gillespie brought Seise with the rest of his household when he moved to San Francisco at the beginning of the Gold Rush. With Seise's support and advice, Gillespie began importing much-needed food supplies and other goods to San Francisco. Gillespie soon faced competition, however, from powerful Chinese companies supplying contract labor both to the mines and to the growing city of San Francisco.[48]

Many ships began crossing the Pacific from Hong Kong to San Francisco, their cramped holds full of workers. The ships also carried familiar food products to keep the workers more or less content with their new circumstances in America. Violence was common against Chinese people in California, whether they worked in mining, fishing, shrimping, farming, railroad construction, or service industries, but at least the food was adequate. By 1852 tens of thousands of Chinese were working in Northern California, many in food services. In the first years of the Gold Rush, Chinese people ran most of the restaurants in San Francisco—although they may have served steaks and pork chops as often as Chinese food.[49]

By the 1860s Chinese businessmen were importing so many bamboo shoots, bean sprouts, noodles, dried oysters, dried mushrooms, and prepared sauces that they paid a half million dollars a year in tariffs.[50] By the end of that decade, one ship alone was carrying the following in its hold:

> 90 packages cassia [cinnamon], 940 packages coffee, from Java and Manila; 192 packages fire-crackers; 30 packages dried fish, cuttle-fish, sharks'-fins, etc.; 400 packages hemp; 116 packages miscellaneous merchandise, lacquered goods, porcelain-ware, and things for which we have no special names; 53 packages medicines; 18 packages opium; 16 packages plants; 20 packages potatoes; 25 packages rattans; 2735 packages rice; 1238 packages sundries—chow-chow [as-

sorted pickles], preserved fruits, salted melon-seeds, dried ducks, pickled duck's eggs, cabbage sprouts in brine, candied citron, dates, dwarf oranges, ginger, smoked oysters, and a hundred other Chinese edibles and table luxuries; 824 packages sugar; 20 packages silks; 203 packages sago and tapioca; 5463 packages tea.[51]

The journalist noted that this cargo was small compared to a Chinese steamer's cargo during the tea season. Americans were fascinated by all the different kinds of foods Chinese people ate.

In this period leading up to the Page Act of 1875, which prevented Chinese women from immigrating to the United States, and the Chinese Exclusion Act of 1882, many Americans treated Chinese people as a threat to society. Tourists, however, were eager to venture into Chinatown, particularly the part along Dupont Street (now Grant Avenue) where street vendors held out elegant tidbits: "leaves . . . rolled into cornucopias to hold a mixture of fig cake, almond, and melon," alongside tall pyramids of buns and sticky sweets. Whole roasted pigs hung from hooks in front of butchers' shops, infusing the street with a rich, smoky smell.[52]

Figure 3.3. Chinese butcher and grocery shop, Chinatown, San Francisco. Courtesy of the San Francisco History Center, San Francisco Public Library.

Groceries offered a wide variety of fresh and cooked meats, fresh and dried fish, poultry, sausage, dried oysters and shrimp, and hashed meats. Markets displayed piles of Shantung cabbage, taro, water chestnuts, lotus seeds, lotus roots, lily bulbs, yam bean tubers, watermelon seeds, winter melons, Chinese turnips, sponge gourds, red and yellow sweet potatoes, and soybeans in many forms. Some of this produce came all the way from China, but increasingly Chinese farmers grew familiar vegetables along the banks of the Sacramento River and delivered them to San Francisco in the mornings.[53]

The standard visit brought tourists to both high-end and low-end eateries, but often just to point and stare, rather than taste and enjoy. Late-nineteenth-century Chinatown was a must-see on the tourist's itinerary. It was also, however, a discomfiting experience, and not just because of the unfamiliar foodstuffs:

A visitor to San Francisco's Chinatown feels as if he had been suddenly transferred to another land. Yet he finds no pagodas with curved eaves and number-

Figure 3.4. Buying fish in the market, Chinatown, San Francisco, 1906. Courtesy of the Library of Congress.

less stories, no oriental palaces . . . but only a series of dingy brick buildings in American style . . . the whole bears, nevertheless, an outlandish look . . . Huge and tiny signboards, all length and no breadth, with vertical inscriptions in red, black or gold, on red or green, white, or black ground, flaunt their moral and florid titles in all directions. . . . Every house in Dupont street . . . bears a number of these signs.[54]

For non-Chinese who wandered into the neighborhood, the oddest part was that the buildings looked completely normal, not like pagodas or palaces, except for their incomprehensible signage.

Chinese restaurants had the same unreadable signs as other shops, but the high-end restaurants also had "Oriental" flair to entice tourists: "they are the only buildings of a true Chinese aspect, forming a most agreeable break in the monotonous dingyness around." The façade of a typical restaurant was painted with a faux-marble design, accented with "arabesque decoration," and topped with an ornate cornice. The brightly painted balconies displayed flowers, tinsel, and colorful Chinese lanterns made of paper or glass.

Many Chinese workers could not afford those elite restaurants. The poorest would simply buy some nuts, a bun, or a bowl of soup from a street vendor when they had earned some coins or could get credit. Or they might pop into the corner grocery: Chinese grocers often had an open kitchen where customers could prepare their small purchases into a meal, topping a bowl of rice with some stir-fried vegetables, pork or fish, and a little lard, sauce, or spice. Customers could also browse through bins of preserved fruits, vegetables, and fish, while dried meat, poultry, bacon, and edible roots were strung from the ceiling.

Chinese laborers making a steady wage might pay a restaurant about nine dollars for a month's board—twice-daily meals when they would sit on benches around a bare plank table, sharing from a common bowl of rice and pork. Each worker had two small bowls, one for tea and one for scooping from the serving dish. With a bowlful of food in one hand and their chop-sticks in the other, they wolfed down their meal and then enjoyed their tea. Salaried employees paid up to twenty dollars a month for meals at a nicer restaurant, offering them a changing selection of dishes and sauces.

The epitome of this fine dining experience, for Chinese businessmen or Americans brave enough to venture into the high-end restaurants, or Chinese employees invited out for a yearly banquet, cost two to ten dollars a head for a grand meal lasting several hours. First, gilt-edged invitations went out, announcing that "a slight repast awaits the light of the guest's presence." Guests gathered in an outer room, enjoying tea and cigars. The

Figure 3.5.   Chinese restaurant on Dupont Street, San Francisco. Courtesy of the San Francisco History Center, San Francisco Public Library.

dining room was "all aglow with lanterns," and the tables were each set for a dozen persons.

Each place setting included a stack of tiny plates, a small bowl, a porcelain spoon, ivory chopsticks, and two large metal cups, one holding a pint of hot tea, the other a pint of rose-scented rice liquor. A soup tureen sat on the table with other appetizers around it, including cucumber and celery salads, pickled duck, ginger, eggs, and melon seeds, as well as salted almonds and other nuts. Many were finger foods: "circular wafers, about two inches in diameter, are often used to envelop mouthfuls of food." The main courses featured such dishes as

> fried shark's fin and grated ham; stewed pigeon with bamboo sprouts; roast suckling pig; boned duck stewed with grated nuts, pearl barley, and mush-rooms; fish sinews with ham; stewed chicken with chestnuts or water-cress; dried oysters boiled; bamboo soup; sponge, omelet, and flower cakes; banana fritters; and birds-nest soup. . . . There are also other dishes which cost up to a dollar a mouthful. . . . Each dish is served cut and minced in quart bowls, many of which are silver-plated and provided with a metal heater in the centre, filled

with coals to keep the food warm. From this the guests help themselves to one mouthful, with the aid of a spoon or chopsticks, and either transfer it directly to the lips or nibble it from the tiny plate before them.

Further courses might include terrapin with onion and water chestnuts; mushrooms with hundred-layer leek; Chinese quail; skewered chicken hearts; rice soup; stewed mutton; roast duck; and mince pies. Next came a succession of fancy dishes, such as "delicate cakes . . . in the form of birds or flowers" and "oranges apparently fresh, but filled with a series of jelly layers of different colors."[55]

This whole world of surprising and delectable treats collapsed suddenly on April 18, 1906, with the great San Francisco earthquake and the subsequent three days of fire and ruin. The Chinese business community soon regrouped and defeated proposals to move Chinatown to a less central neighborhood. To build support among San Franciscans, Chinese leaders offered to feed earthquake refugees of any ethnic background who came to Chinatown's relief kitchens. Chinese merchants seized the opportunity to change the city's view of Chinatown, hiring white architects who added an "Oriental" aesthetic to all the buildings and street furniture. The district began to look more like an exotic stage set representing China. Restaurants catered to San Franciscans seeking out the newest food trend: chop suey joints. Simultaneously, the actual Chinese population in Chinatown began to contract, from twenty-five thousand in 1890 to fewer than ten thousand by World War I.[56]

In the 1920s Chinese restaurants began to pop up outside Chinatown: south of Market or near City Hall. The New Shanghai Terrace Bowl opened in 1927 and appealed to both Chinese and non-Chinese San Franciscans, distributing coupons in popular department stores. Times were hard during the Depression, when people often had barely enough rice for their families and had trouble scrounging up a dime for some *sung* (toppings for one's rice). During World War II, American attitudes toward the Chinese community grew warmer in acknowledgment of the alliance between China and the United States. Chinese Americans in San Francisco started the China War Relief Association, organizing "Bowl of Rice" events as fundraisers to relieve suffering in Japanese-occupied China and help the war refugees. These community events drew in white Americans as well and popularized Chinatown as a vibrant part of the city. In 1940 Johnny Kan opened his first restaurant, Cathay House; in 1953 he opened Kan's, having weaned San Franciscans away from a chop-suey understanding of Cantonese food.

In 1943 the Chinese Exclusion Act was finally overturned, and Chinese living in the United States could apply for naturalization. Even as many

Kan's

708 GRANT AVENUE · CHINATOWN · SAN FRANCISCO

Kan's Nine Course Peking Duck Dinner — $35.00
Serves Eight Persons

Over Eight Persons Add $4.00 Per Person
(Two Days Advance Notice Required)
MELON CUP SOUP (Doong Gwa Joong) in Season Only
"The Piece de Resistance"—PEKING DUCK (Ka Law Opp) Served with Hot Steamed 9 Layer Buns
SWEET AND SOUR PINEAPPLE PORK (Goo Lo Yuke)   GOURMET VEGETABLES (Gah Ming Yeong)
CHICKEN IN PARCHMENT (Gee Bow Gai)   HOP TO GAI KOW (Walnut Chicken)
PRECIOUS FLOWER EGG (Gwai Fah Don)   YEONG JO FRIED RICE (Yeong Jo Chow Fan)
DESSERT—ASSORTED CHINESE TEA CAKES   KAN'S EXOTIC ALMOND EYE or CHILLED LICHEES
CHOICE OF—OOLONG, JASMINE or HOONG TEA

Figure 3.6. Kan's menu, 1953.

Chinese assimilated, the variety of Chinese restaurants and their clientele endured. In 1960 sociologist Rose Hum Lee noted that in San Francisco Chinese food was eaten in tourist restaurants, neighborhood restaurants across the city and suburbs, popular weekend eateries in Chinatown, old-style banquet halls, and take-out places. High-end Chinese food was also served alongside other cuisines in American restaurants promoting themselves as exotic. Apparently Chinese food could still be "exotic" more than one hundred years after the first Chinese restaurants opened in San Francisco.[57]

Japanese

The Japanese were fashionable in San Francisco before they were even present in the city. Japanese bazaars sold curios, lacquered boxes, fans, and vases, and almost all grocery stores had Japanese teas. Commodore Matthew Perry demonstrated American might to Japan in 1852; fifteen years later, Japan's ruling shogun had been ousted. San Francisco traders repurposed his diplomatic title, "Taikun," to refer to high-quality "Tycoon" tea. Japan's new Meiji government rushed to industrialize and open up to the world.

When Japanese immigrants started arriving in San Francisco in the 1870s, many Americans appreciated them as domestic servants. They were more formal than Chinese immigrants, and, coming from Meiji-era Japan by way of Hawaii, they were also more westernized—with Western clothes and hairstyles, and some knowledge of Western dishes and Western methods of food preparation. Those who were neither students nor domestics established small shops, boardinghouses, and restaurants for other Japanese immigrants in or around Chinatown or near the docks. Some distributed Japanese artisanal goods, including handmade earthenware bottles of "Mikado sauce" (an early *tonkatsu* sauce, related to barbecue sauce). Others went into small-scale production of soy products. For the most part, they tried hard to avoid competing with Westerners; they presented themselves as the good Asian minority.

That lasted until the early 1890s. As the United States hovered on the brink of an economic depression, the White Cooks and Waiters' Union gained in strength and began to strong-arm other workers into eating only at union restaurants. Japanese restaurant owners applied for union status, but the union would not accept Asians and shut out all Japanese restaurants. Nor could Japanese restaurant owners buy effective protection from union boycotts and violence—they tried bribing union officials but lost their money to no purpose as the intimidation continued.[58]

At the same time, the expanding Japanese population was beginning to compete for agricultural jobs with white farmworkers. In August 1893 anti-Japanese rhetoric spread rapidly in Napa, where Japanese were helping to gather and dry the prune harvest. The ugly sentiments barely slowed the trend of Japanese working in Bay Area orchards and other agriculture. By the time of the Russo-Japanese war in 1905, many Californians were worried about their jobs and also about a possible military threat across the Pacific. A *Chronicle* article complained that Japanese men were taking kitchen and waitstaff jobs in non-Japanese restaurants. That same year the Exclusion League of Alameda County held a meeting excoriating "coolie labor" and complaining about Japanese running restaurants, saloons, vegetable stores, and gardens in Oakland, as well as a brewery in Berkeley. Others focused on the region's orchards, warning farmers about their Japanese workers: "Mr. Fruit-raiser will find in twenty or thirty years' time that the Japanese hold the orchards."[59]

After the earthquake in 1906, many Japanese were priced out of Chinatown and moved to a cheaper part of town, the South Park area south of Market Street. Japanese immigrants disembarked nearby and found their way to South Park's Japanese boardinghouses, as did Japanese farmworkers waiting for the spring planting to begin. Japanese restaurants sprang up along Third Street and Brannan, and beyond. But soon the White Cooks and Waiters' Union began targeting these restaurants for harassment. The San Francisco branch of the Exclusion League insisted that "wage-earners are to be cautioned against the danger to their health and that of their families in eating berries picked and packed by unclean and unhealthy Asiatics."[60]

That same fall, however, restaurant critic Clarence E. Edwords published in the *Chronicle* a quite favorable impression of a "genuine Japanese dinner" he enjoyed in an upscale restaurant tucked away in an unobtrusive house on O'Farrell Street:

> For daintiness and artistic coloring no dinner could excel it. For deliciousness of food and at the same time mystery in externals it was a constant surprise. That which appeared to be a frosted cake of delicate pink body and snow-white coating proved to be a slice of pressed sea bass with lobster as the frosting. A delicious looking fruit cake changed to a combination of fish and chicken when it was placed in the mouth.[61]

He revisited this dinner in his 1914 book, *Bohemian San Francisco*, revealing the rest of the meal:

> Immature vegetables, served in a sort of sauté. These were sprouting beans, lentils, peas and a number of others with which we were unfamiliar. . . . One of the women came in bearing aloft a large silver tray on which reposed a mammoth crayfish, or California lobster. This appeared to be covered with shredded coconut, and when it was placed before the host for serving he was at a loss, for no previous experience told him what to do. It developed that the shredded mass on top was the meat of the lobster which had been removed leaving the shell-fish in perfect form. It was served cold, with a peculiar sauce.
>
> Now followed the piece de resistance. A tub of water was brought in and in this was swimming a live fish, apparently of the carp family. After being on view for a few minutes it was removed and soon the handmaidens appeared with thinly sliced raw fish, served with soy sauce. Ordinarily one can imagine nothing more repulsive than a dish of raw fish, but we were tempted and did eat, and found it most delicious, delicate, and with a flavor of raw oysters.[62]

The old interest in Japanese style had not vanished from San Francisco—Edwords was fascinated by this chef's playful aesthetics.

Other Japanese moved west to the Western Addition neighborhood and formed a *Nihonmachi* (Japantown) to the north of Geary Boulevard. San Francisco's Japantown is the oldest in the continental United States. The neighborhood had Japanese grocery stores; a beer parlor; a confectionery named Benkyodo that sold *mochi* (sweet, chewy rice cakes), *manju* (bean-paste buns), and Japanese fortune cookies; the Uoki Sakai seafood shop; a vegetable market; a lively restaurant with great *sake*; and another restaurant specializing in eel. Japanese families tried to live normal American lives, even as they faced growing mistrust from the community around them. Japanese who emigrated from Japan could not become citizens or own property in the United States, but they held to the belief that their children would have those rights. Sadly, their children instead faced the U.S. government's Japanese internment program during World War II. While the Japanese were eating lima beans, canned food, and Jell-O at the Tanforan Assembly Center in San Bruno, and then hot dogs, Spam, and potatoes at the Topaz War Relocation Center in Utah, other ethnic groups came to the Western Addition to take over the emptied-out neighborhood.[63]

After the war, returning Japanese residents tried to settle again in Japantown but faced several decades of redevelopment efforts. Through eminent domain the city razed three square blocks in the center of the neighborhood, ousting many postwar restaurants such as the popular Matsuya restaurant, which moved to Bush Street in 1968, and then to a tiny space in Noe Valley. (Matsuya made the *Chronicle*'s list of top restaurants in 1996 after longtime owner Fusae Fukuda Ponne charmed food editor Michael Bauer with her sushi and her wisecracks.)

Back in Japantown, Japanese investors built shopping malls aimed primarily at tourists. By the late 1980s, however, the Kintetsu Mall also included the locally popular Maruwa food market (taken over by the Nijiya chain in 2004). Trendy restaurants in Kintetsu included Isobune, a sushi bar that introduced Americans to the pleasure of watching their dinner options float by on tiny boats. Across the street was Sanppo, 1702 Post Street, which still serves savory noodles, *gyoza* (dumplings), and *nasu hasamiyaki* (ginger-marinated beef broiled with Japanese eggplant). Korean restaurants soon moved into Japantown as well, including Korea House and its offshoot, New Korea House, known for their grilled meats and *chap chae* (clear sweet potato noodles).

Italians

At the time of the Gold Rush, merchants and adventurers from the Genoa region had been working in South America. Many of them took the news of gold as a call to head north, and they also sent word to Italy to encourage their friends and family to join them. But the first Genoese immigrants had a head start in cross-cultural business dealings and ultimately played a large role in San Francisco's history.[64]

One of them was Domenico ("Domingo") Ghirardelli, who came to Northern California from Peru in 1849. He first opened a grocery store selling fresh olives and olive oil, cognac, muscatel wine, Genoese wine, Florentine superior sausages, sardines, and assorted vermicelli, as well as products from South America such as white Peruvian sugar, fresh Chile flour, and—of course—chocolate. He then opened a coffeehouse, offering customers hot chocolate as well. By 1853 he had brought his wife, Carmen, and their children from Peru to San Francisco and established "Mrs. Ghirardelli & Co.," a manufactory of syrups, ground coffee, and chocolate liquor. Domingo decided to specialize in chocolate, importing industrial equipment to grind and roast the beans, and then to mix and aerate the liquid chocolate. For almost a century Ghirardelli was the largest chocolate producer on the West Coast. The company left its mark on the city with its famous clock tower and enormous electrified sign; it also left its mark on generations of San Franciscans who loved Ghirardelli chocolate as a drink, an ingredient, or a tasty treat. That bond solidified in the 1960s when the company's dated factory became the historic Ghirardelli Square attraction.

One hundred years earlier, other Genoese immigrants became popular for their pasta. In the 1850s Brignardello and Macchiavello were known for their steam-powered factory putting out vermicelli and macaroni: "Their goods are judged to be superior to any imported from Italy; they possess a freshness and beauty of careful finish."[65] Grocers sold these products to

Figure 3.7. Ghirardelli Square, Beach and Larkin Streets. Courtesy of the San Francisco History Center, San Francisco Public Library.

Italian and French eateries and to home cooks, but everyone agreed that non-Italians overcooked their pasta: "We have an insensate way of tumbling a handful into a basin of soup, which comes out a pasty stuff, good for nothing."[66] Italian cooks knew how to make pasta, but there were not yet very many of them in San Francisco. Nevertheless, by the mid-1850s, an area on Dupont Street was known as Little Italy; the community then moved toward Telegraph Hill and North Beach in the 1860s. The Italians never isolated themselves from other ethnicities, however, and even North Beach was not an Italian enclave. As more Italian immigrants arrived, they began to make their homes in other neighborhoods, such as Portola, Potrero Hill, Noe Valley, and the Mission.[67]

In the 1870s San Francisco's Italians were far outnumbered by the Irish and Chinese. Then, in 1875, the California Immigration Union began publicizing the idea that California was "The Ideal Italy of the World." Promoting the similar climate and landscape proved an effective way to reach out to Italians. Most Italian immigrants had worked back home as farmers or fishermen; many continued to do such work in America, although some branched out into other fields. In 1886 the Italian bureau of statistics reported that about seven thousand Italians were in San Francisco, many of whom worked in various food industries. The bureau counted five Italian firms in wholesale imports, twenty-two food retailers, ten fruit wholesalers, seventy-three fruit and vegetable shops or vendors, eighty-five vintners or wine dealers, forty-eight wine and spirits shops, fourteen butchers, twelve *trattorie*, seven pasta manufacturers, four confectionery makers, seven bakers, three delicatessens, two dairies, and one dealer in chocolate and coffee.[68] Italian truck farmers brought their vegetables to Colombo Market, the new covered hall at Davis and Pacific Streets: "It is delightful to see the bustle of the beautiful two-horse wagons marked with their owners' names."[69] These market gardeners knew they would have Italian customers for their bell peppers, eggplants, zucchini, fava beans, artichokes, and plum tomatoes, and before long they had non-Italians clamoring for their delicious produce as well.

By 1900 the Italian community was growing quickly, adding twenty thousand new immigrants over the next two decades—until the Immigration Act of 1924 blocked most immigration by southern and eastern Europeans, as well as by Asians. About 70 percent of the immigrants were from Genoa, Tuscany, and other regions of northern Italy, and many of them focused on fishing. Italian San Franciscans reminisce about heading to Fisherman's Wharf to pick up fresh seafood from a stand, or catch it right off the pier. (Before 1900 they would have headed to "Italian Harbor," off the Filbert Street Wharf in North Beach.) No need to sit down in a restaurant to enjoy a crab feast—just take your catch to a park and cook it over an open fire. Italians maintained their hold on the Fisherman's Wharf seafood trade well into the twentieth century.

Italian specialty stores also had their aficionados. Delicatessens hung sausages encased in tinfoil from the ceiling, dazzling the eyes as they twisted; the owners stacked Italian eels in tins next to candies in brilliant wrappers to attract the eye. Pasta emporiums displayed yards and yards of tagliarini, vermicelli, lasagna, and reginini (similar to fettuccine) in addition to about seventy-five other kinds of pasta. Italian women would bring in their own homemade fillings to be stuffed into raviolis at stores such as the Panama Canal Ravioli Factory and its nearby competition, the Canal Exposition Ravioli Factory. Bakeries offered Italian-style loaves—"big crescents and the hard nubs of bread that every traveler in Italy has sighed over." Wine dealers pressed with their feet "that execrable vin d'uva so dear to the fisherman's heart."[70] In Italian neighborhoods, tomatoes were laid to dry in many backyards next to rows of basil, rosemary, thyme, oregano, marjoram, and fennel; strings of garlic dried on rooftops; and Italian fishermen sat mending their nets while goats wandered up and down the hilly streets. On Saturdays, Italian housewives visited each shop, buying fresh ingredients for the next day's culinary extravaganza; on Sundays, each household gathered together for a two-hour midday meal that communicated the role of food in holding a family together. Once children moved to the suburbs, they would still return to their original neighborhood, their own personal Little Italy, for Sunday dinners, or at least for special events such as baptisms, weddings, funerals, and feasts of Italian patron saints.

Irish

In the 1880s the Irish were the largest foreign group in San Francisco, totaling about thirty thousand people. Many had been in the United States for a while and owned real estate; some even had land grants dating from the Mexican era and ran their own ranchos north of the Golden Gate. Edward and Margaret Mitchell had a commercial dairy in Noe Valley from about 1860 to the 1890s; half a century later, their grandsons opened Mitchell's Ice Cream a few blocks away on Twenty-ninth Street. John D. Daly did not start off with money but came to San Francisco as a boy of thirteen and worked for years in San Mateo dairies. Daly acquired his own dairy in 1868 and began distributing milk and eggs in the Mission district; he played a large role in San Francisco's reconstruction after 1906 and, in 1911, Daly City was named after him.

In the late nineteenth century, more poor Irish arrived every year, and many of them ended up in San Francisco households as servants and cooks. Food historian Hasia Diner has shown that the Irish in America did not talk about their foodways the way other immigrant communities did. Memories of the potato famine were still too sharp, and the Irish had few positive food memories on which to build a cultural tradition.[71] Every year the Irish-American societies, the Sons of Erin, the Knights of St. Patrick, the Irish Confederation, and so on, would celebrate St. Patrick's Day in style, with dancing and sporting competitions, as well as with convivial banquets, speeches, and many toasts. Newspaper write-ups of these events simply mentioned the "excellent repast" or the "sumptuous banquet." Once, in 1895, someone did reveal the menu to a reporter, and it consisted of oysters on the half shell, "Soup a la Reine," lobster with mayonnaise, English sole with hollandaise sauce and new potatoes, filet of beef with fresh mushrooms, croquettes of sweetbreads, asparagus *en branches*, cauliflower au gratin, Roman punch, "Spring Chicken a la Cassarole," salad, and tutti-frutti ice cream with assorted cream cakes, bonbons, fruits, and cheeses for dessert. To feel like an event, the food had to have a glamorous French aura. The Irish in San Francisco performed their Irishness to each other through dance and games, and they demonstrated their power to politicians by throwing elegant French banquets accompanied by long, rousing speeches. Irish food was not part of that performance. In 1922 Irish ingredients finally appeared at a St. Patrick's Day banquet, but just for display: the organizing committee had decorated an illuminated fountain in the middle of the Palace Hotel's Palm Court with festive piles of potatoes, red and green cabbages, turnips, parsnips, cauliflower, onions, and lettuce.[72]

At occasional moments, Irish culinary pride did emerge. In 1891 when the Women's Educational and Industrial Union organized its "Feast of Nations" fundraiser, the board asked an Irish cook to oversee the cosmopolitan dinner. She managed the chaos in the kitchen as the lady benefactors (and two men) demonstrated various culinary traditions. The "big lady from Erin" not only supervised the production of sixty-eight different dishes and almost as many cooks, but also managed to produce two Irish dishes herself—cabbage and Colcannon.[73] Thirty years later, the *Chronicle* offered a recipe:

Colcannon: Two cups parsnips, 2 cups turnips (white), 2 cups cabbage, 2 cups onions, 2 cups potatoes, salt and pepper, sausage meat. The vegetables should be pared and cut in small pieces, put on to boil in the order given, allowing five minutes . . . before adding the next. When potatoes are soft, drain carefully, mash all together, season to taste, mound on a hot platter, surround with a ring of nicely browned rolls of sausage meat and serve hot. . . . This dish always appears at Halloween with a plain gold ring concealed in its depths, which tradition says will appear on the plate of the person who will be the next married.[74]

Other symbols of Irish culture appealed more to San Franciscans. At an 1898 "Irish Fair" attended by then mayor James Phelan, the food committee prepared "old-fashioned Irish stews, the same as served in Dublin."[75] Irish bread and scones were also popular. Over time, Americans began to view soda bread, corned beef, and cabbage as age-old Irish traditions. San Francisco inspector William J. Delaney encouraged that belief by throwing a corned beef and cabbage banquet at his elegant home on Clay Street. Attendees said the banquet "should go down in the epicurean history of San Francisco as one of the most successful dinner affairs."[76]

For the most part, however, Irish Americans ate like other Americans. Domestic service provided a way for Irish women to learn how to make new dishes. Their employers were pleased that the Irish, unlike other household cooks, did not tend to fall back on their own cultural traditions when hurrying to put dinner on the table:

> There are more Irish cooks in the employ of Jewish people than any other race. Some of these, usually young girls, become such mistresses of the mysteries of Hebrew cookery that they leave little to be asked for from the heads of the house, who are prone to grumble at "foreign" introductions on their daily bills of fare.[77]

Not surprisingly, employers did find other aspects to grumble about; they claimed it was particularly difficult to find servants with pleasant attitudes. An editorial in the *Chronicle* proposed that employers solve that problem by making their servants' lives more pleasant:

> You would never guess the one thing I have done which seems to have given the greatest satisfaction and . . . had the best results. My one servant now is an Irish woman of the better class, and, as you may imagine, she is a great tea drinker. . . . [However,] she did not care for tea if she had to drink it from a thick cup. In other words, she is something of a connoisseur. . . . I have now fitted her out with a pretty set of thin china for her own use, and you have no idea how pleased she is. There are two cups and saucers, several plates and a pretty little creamer and sugar bowl, and she can have friends visit her occasionally.[78]

Normally, Irish women were on their feet all day, starting early in the morning when they stopped by the wholesale market to pick through the scraps for still edible carrots, turnips, dill, summer savory, and lettuce to bring back for the stockpot. One of the few possible pleasures in an Irish woman's day was to sit in the kitchen with a cup of tea and chat with a friend who had dropped in.

Over the course of the twentieth century, it became normal for Irish Americans to eat what they saw as "American" food. Michael Killelea remembers shelling peas on the porch with his sister, to accompany the regular Sunday evening leg of lamb and baked potatoes at their house in Berkeley. But he had even fonder memories of ordering a Coke or an ice cream soda at the Woolworth's counter, or going to McCurdy's on Shattuck to order a hamburger and a milkshake. Sometimes his whole family would head to Kress's for dinner, eating meatloaf and potatoes with gravy, again at the counter.[79] This harked back to the way Irish laborers used to eat when they had just arrived in San Francisco and had no more than a dime to spend on a meal:

> The perpendicular refreshment dodge . . . dispenses with tables, waiters, etc., the customers sitting on high stools or standing at the bar and disposing of chops, steaks and Irish stews with a rapidity fearful to witness.[80]

It is hard to overstate the enduring appeal of cheap, satisfying food to large Irish families living in and around San Francisco.

Jews

One of the best-connected Jews in early-twentieth-century San Francisco was an Irish immigrant, Albert "Micky" Bender. His mother and his rabbi father were both of German origin, but Bender himself was born and raised in Dublin. He came to San Francisco in 1881, did well selling insurance, and became one of the most prominent patrons of the arts. He gave delightful dinners for artists and their benefactors, where the food was "very tasty and very handsome." However, Ansel Adams complained that because Bender did not drink himself, or even eat much beyond scrambled eggs and toast, he did not notice when the staff mixed the scotch with ginger ale instead of soda, or forgot to stock ice for a party.[81]

Earlier Jewish immigrants on the West Coast were likewise of German origin, taking a wide variety of paths on their way to California. During the Gold Rush, San Francisco had its own matzo bakers and kosher boardinghouses. Sometimes they ran into conflicts with others for working seven days a week, "neither observing the Jewish Sabbath nor our own" and supposedly using their religion "as a cloak for avarice." Commentators alleged that Jewish men worked every day except Rosh Hashanah and Yom Kippur.[82] Many San Francisco Jews were well established by the end of the nineteenth century. They displayed a diversity of practices: from concern with kosher law to participation in secular Christmas festivities, including Christmas cookies. The prominent Haas family belonged to Temple Emanu-El but held Christmas parties every year and did not keep a kosher home: "We might dine on gorgeously glazed hams . . . Bacon sizzled in the kitchen." Even those who kept kosher at home were often happy to enjoy oysters or lobster when out on the town.[83]

Harriet Levy's well-to-do Polish-Jewish family kept kosher, and she remembered sharing a conspiratorial glance with their Irish cook, Maggie Doyle, whenever Harriet accidentally cut the butter with a meat knife. Harriet loved spending time in the kitchen, watching Maggie interact with the deliverymen:

> Twice a day the tradesman or his emissary knocked at the kitchen door. In the morning he took the order, in the afternoon he delivered it. The grocer, the baker . . . the fish man, the chicken man, the butcher boy, came twice a day proffering a salty bit of conversation or a flashing glance of fire. Bent over the sink, Maggie Doyle aimed a shaft of repartee over her left shoulder. . . .
>
> The men brought the wind and the rain into the house on their faces and hair, and they bore the smell of fish and vegetable and meat. . . . When Christmas came, the tradesmen brought presents . . . in kind. . . . The most welcome offering was that presented by the baker, a huge yellow cupcake hidden under a coat of heavy white icing with Merry Christmas outlined upon it in crinkly pink sugar letters. . . .
>
> The baker was German; the fish man, Italian; the grocer, a Jew; the butcher, Irish[;] . . . Chung Lung was fruit and fishman. He carried the combined stock, suspended in two huge wicker baskets balanced upon a long pole across his shoulders. He did not come to the kitchen door, but remained outside in the alley. We heard the back doorbell and hurried down, Mother, Maggie, and I. The uncertainty of the contents of the baskets—today only apples and cauliflower; tomorrow, cherries and corn; today, shiny silver smelts; tomorrow, red shrimps with beards and black-beaded eyes—made a delight of his coming. . . . Chung [taught] me how to twist the tail of the shrimp to make the meat pop out, unbroken.[84]

Maggie had experience cooking for a Jewish family even before the Levys and took their expectations in stride. She kept milk dishes separate from meat dishes and both apart from the Passover dishes; she prepared holiday traditions such as beet soup or "epicurean dishes made of Matzoth meal"; and if company were coming she might make lobster salad or shrimp canapés. On the eve of Yom Kippur, Harriet's father would perform an ancient ritual, whirling live chickens over the bowed heads of his daughters, feathers flying everywhere, then a duck over his wife's head and a drake over his own, offer-

Figure 3.8. Chinese vegetable peddler, San Francisco. Courtesy of the San Francisco History Center, San Francisco Public Library.

ing up the poultry as a sacrifice in place of his family. At the end of the Day of Atonement, Maggie cooked up the duck, and "with sufficient gravy, we were able to eat it." Keeping the old traditions did not interfere with trying new tastes, as when Harriet was out walking on a Saturday evening with her father and bought hot tamales from a street vendor.

Eastern European Jews began arriving in large numbers but found much less of a welcome in San Francisco than in New York, Chicago, Philadelphia, or Los Angeles. In 1915 about ten thousand Jews were in San Francisco, of whom only one-third spoke Yiddish or had parents who did.[85] Before 1906 most of San Francisco's Eastern European and Russian Jews lived south of Market. This was a cheap part of town—"south of the slot," as it was called, because of the cable cars on Market Street whose cables went through a slot. But it was not poor in culture. Jewish blocks had kosher butchers, dairies, and bakeries, as well as fruit markets and candy stores to delight children on their way home from school. Jewish peddlers came by with horse-drawn carts, hawking produce along with ice for your icebox. Observers noticed differences between some of the "hyper-orthodox" new arrivals and earlier Jewish immigrants: "Fancy a San Francisco restaurant-fed Jew insisting that his meat be freshly killed [by] kosher methods. . . . He bolts his pork and sausage . . . and it is all over and under with the oyster or lobster or crab coming within his capacious reach."[86]

After the earthquake, most of the Jews "south of the slot" moved to the Fillmore district, bringing with them kosher shops and bakeries making familiar loaves of pumpernickel or rye. There were five kosher meat markets, four Jewish bakeries, three Jewish restaurants, another three Jewish delicatessens, and a Jewish liquor store, all clumped in a small neighborhood framed by Fillmore, Buchanan, McAllister, and Golden Gate Avenue. The scent of garlic-dill pickles and smoked herring permeated Jewish childhoods in the Fillmore district, along with the sound of squawking chickens. In the 1920s children visited the Cat & Fiddle candy shop for a malted milk or to crank the taffy machine. Or they crossed Fillmore to go to Boston Lunch, with a whole row of waffle irons placed invitingly in the window. On Saturday nights, many shops opened after sundown and stayed open until midnight, creating a festive atmosphere where housewives might buy picnic supplies for the next day. Among the most enduring institutions were bakeries such as Langendorf's and Waxman's, Goldenrath's delicatessen, Shenson's Kosher Meat Market, Diller's kosher restaurant, and Sosnick's Market. For Rosh Hashanah, Langendorf's would display huge challah loaves, some topped with colorful candy sprinkles. Schubert's Bakery offered flaky macaroon tarts, chocolate éclairs (six for a quarter), and burnt almond cake with piles of whipped cream. Jewish stores in the Fillmore sat next to shops and restaurants run by Mexicans, Greeks, Italians, and many other ethnicities, so there was also plenty of opportunity to try new tastes. One popular option was to head a few blocks north to the Wagon, on Post, where ten cents would buy a fried-egg sandwich, or "the best hamburger sandwiches in town."[87]

After World War II, many Jews moved out of the Fillmore to other parts of the city, including the Richmond district, home to many of San Francisco's Russians. There, Jewish entrepreneurs built new culinary touchstones, including Shenson's Delicatessen and Sid Chassy's House of Bagels, facing each other across Geary Boulevard. Jewish families over the decades established favorite food routines in their new neighborhoods. For some, those traditions included visits to David's Delicatessen, still located on Geary over by Union Square and the theater district, or to Saul's, across the Bay in Berkeley.

Russians

A few Russians had come to San Francisco before it was even called San Francisco. In 1806 Nikolai Petrovich Rezanov came to the Presidio seeking emergency food supplies for Russian settlers in Alaska. In 1812 Russians established an outpost to the north of the Bay Area, to supply food for fur-trading communities even further north. From that outpost, known as Fort

Ross, they engaged in trade with Yerba Buena, the future San Francisco. Russians who died at Yerba Buena were buried at the top of what became known as Russian Hill, with crosses engraved in Cyrillic to mark their graves. Russians did not, however, make up a large part of the Gold Rush community in San Francisco, and the few who were there did not live on Russian Hill. In 1864 Eastern Orthodox Russians founded a parish with the city's Greeks and Serbs; in 1871 they established a Russian school for their children; and in January 1888 a *Chronicle* reporter noted Russian children's parties around the city marking the Julian calendar's new year. Molokans, Baptists, and Jews also came to San Francisco, seeking freedom from religious persecution in Tsarist Russia. Vascovich's restaurant at Stockton and Clay provided a place for Russian men to congregate, though it attracted an unruly clientele. Molokans probably did not eat there—these spiritual Christians were teetotalers, who got their name from their practice of drinking milk (*moloko*) during traditional Russian Orthodox fasting days.[88]

In 1917 the Russian Revolution sent many more refugees to the United States, and to San Francisco in particular. These "White" Russians were aristocrats, officials in the Imperial Army, merchants, priests, and others who opposed the Bolsheviks. Valentina Alekseevna Vernon, who ran the Russian Tea Room, remembered those days:

> At this time the Russians were very close to each other. Most of them lived near the Fillmore district, but at this time Fillmore was a very nice street. . . . There were kind of arches and big lamps, with electric globes. It was really very pretty, and all the stores were Jewish stores, and those Jews were very, very nice. They were mostly Russian Jews, and spoke Russian and helped the Russians to settle in the Fillmore district. They had come before the Revolution; they had stores, different kinds of groceries and so on.[89]

So the new Russian immigrants were able to re-create some of their foodways and traditions from the old country, in the Fillmore district and in the Richmond district to the west. They would visit each other's homes, paying their respects over vodka, tea, and caviar. At home, Russian mothers would make borscht, *ouka* (fish soup), stuffed cabbage, fish balls, *piroshki* (meat-filled buns), *lapsha* (noodles, for chicken soup), and *pelmeny* (meat-filled dumplings), as well as special Easter treats such as *kulich* (a tall, saffron-flavored loaf) and *paskha* (sweetened cheese molded into a pyramid).

Russian entrepreneurs also began opening restaurants for the broader tourist market. The Russian Tea Room opened on Russian Hill during Prohibition, and the staff used to have to keep an eye on the customers: "they brought their liquor in those flat bottles and the ladies had them in their purses . . . and they would pour the water out and pour in the liquor from their flasks." After the restaurant moved to Sutter Street, Vernon struggled to keep it open during the Depression—it was only possible because her chefs, her waitstaff, and her Filipino kitchen help accepted deep cuts to their wages. At the same time, Vernon received very generous credit from her suppliers, including Mr. Waxman, the Jewish baker, "who made excellent Russian bread, I have to admit."

In the 1940s the beautiful Balalaika Restaurant on Bush served up piles of caviar, potatoes baked with cream, chicken Kiev, chicken stroganoff, and "shashlyk Caucasian served on a flaming sword" (skewered lamb), all accompanied by balalaika music.[90] Similar dishes made their appearance at Troika, a restaurant on Powell. Starting in the 1960s, tourists and locals alike enjoyed the excellent food and colorful Russian artwork decorating Luchina on Clement and the gaudy Renaissance Restaurant on Geary. They delighted in *piroshki* on Clement, either at the Miniature Restaurant and Bakery or at the Park Presidio Restaurant and Bakery, which also offered *syrniki* (cottage cheese pancakes), kasha, sausages, and cabbage rolls.

By 1979, however, when Archil's opened off Union Street near Fillmore, it broke new ground by moving away from the touristy environments offered by the established restaurants, offering innovative, reasonably priced Russian cuisine, with no kitsch, no samovars, no balalaikas, and no flaming skewers. Instead, Archil's offered:

> Tender soused herring topped with scallions and cilantro. . . . Fish *kotlety* . . . like a quenelle that's been browned, its flesh moist and tender. . . . Stout peasant *golubsty*, cabbage leaves rolled around a heavy filling of rice and ground beef and relieved by a smooth, sweet-and-sour tomato sauce.[91]

A decade before the fall of the Berlin Wall, already times were changing when it came to Russian cuisine in far-off San Francisco.

Filipinos

Before the Spanish American war of 1898 and the subsequent American annexation of the Philippines, only the occasional Filipino visitors made their way to San Francisco. In late November 1902, however, a journalist noted Filipinos on Kearny Street—"as much at home as in their native land"—haggling with a peddler over a slice of fish or eating a cheap meal in a Chinese restaurant.[92] By 1918 a couple of Filipino brothers from the Visayan islands had opened Santa Maria Restaurant on Jackson near Kearny, and a burst of immigration from the Philippines was beginning. Filipino laborers first went to work on Hawaiian plantations, before making their way to San Francisco in hopes of a better life. They mostly tended crops such as asparagus, celery, lettuce, and beets, traveling north to Seattle or Alaska during the winter canning season. Between seasons, these single men rented rooms in San Francisco hotels in the Kearny Street area, called "Filipino town." Before 1968 it was not yet known as "Little Manila"—indeed, few of the immigrants were from Manila. Most were from rural parts of the Philippines and spoke a variety of dialects; in America, they learned Tagalog to communicate with each other. They also began to think of their community as enjoying a cohesive Filipino cuisine instead of focusing on the regional background of particular dishes.[93]

Chinese and Japanese groceries catered to this new market by adding products from the Philippines into their regular shipments from Hong Kong. They would get in bottles of *bagoong* (fish or shrimp paste) and coconut milk; canned *lanzone* fruit (a yellow berry with a sweetish sour taste); nut pastes (from *casuy* and *pili* nuts); and some salted or smoked fish. Over time, Filipino families settled in other neighborhoods, south of Market, in the Western Addition, or down in Daly City, which became known as the *adobo* capital of the United States. These families began to seek out familiar Filipino products, to please each other and their neighbors with tastes from home.[94]

They grew *saluyot* (somewhat mucilaginous greens) and *tanglad* (lemon grass, for stews and barbecues). They aged duck eggs in salty brine for weeks at a time. They prepared favorite dishes such as *pinakbet* (boiled pork, *bagoong*, and a host of vegetables—bitter melon, eggplant, tomato, okra, string beans, chili peppers, winged beans, and moringa pods); *kare-kare* (oxtail in peanut sauce with banana flower buds, eggplant, string beans, and Chinese cabbage); *sinigang* (stew, made sour with tamarind or lemon juice); and Filipino *adobo* (pork or chicken, slow-cooked in vinegar and soy sauce, then browned). When they could not find certain ingredients in the Bay Area, they substituted local products; these common substitutions helped build the idea of a shared Filipino American cuisine.[95]

In 1969–1970 when Herminia Quimpo Meñez did ethnographic research among California Filipino communities, she found more culinary similarities than differences:

Foodways are basically alike in different Filipino communities in California. The retention of regionalisms is minimal. . . . Also, foods have lost their socio-economic valuations. . . . Festive foods in the Philippines like *lechon* (pork barbecue) and fried chicken might be daily fare in California. . . . The dichotomy is made instead between Filipino and American cookery. Questions as to whether the menu is to be American or Filipino are vitally important in [planning] social gatherings.[96]

Thirty years later, when Joaquin Lucero Gonzalez made a similar investigation into Filipino foodways in Daly City and downtown San Francisco (the parish of St. Patrick's Church), he found Filipino Americans delighting in many of the same dishes, but the bright line between American and Filipino food was no longer so visible.[97]

Home cooking for community events began to display a range of culinary influences. Regional dishes requiring special ingredients made a big impression, and people would remember who made a great *tinola* (chicken soup with green papayas, ginger, and peppercorns), or who arranged for the *lechon* (roast suckling pig, stuffed with tamarind leaves). Yet people also brought typical American dishes such as fried chicken or spaghetti with meatballs. Guests enjoyed Filipino appetizers, such as *chicharon* (pork crackling), *dilis* (dried fish), *kinilaw* (raw fish marinated in vinegar), and *lumpia* (egg rolls), but were also happy to have hot dogs, without seeing any clash of cuisines.

The proximity and convenience of cheap Chinese and Japanese restaurants presented an obstacle for Filipino entrepreneurs hoping to open their own restaurants. By the 1930s, however, the Santa Maria restaurant had successors. The Luzon Restaurant offered a jukebox and "home cooking," and the Rizal Café offered "sanitary cooking and courteous service," as well as "private booths for ladies." New Luneta Café had a pool hall cum barbershop at the front of the house; the café where Mrs. Basconcillo prepared chicken *adobo*, *pancit* (noodles), and rice was in the back of the house. Even further inside, Mr. Basconcillo ran a gambling hall, which brought in more money than the other businesses combined. In the 1940s more than two dozen Filipino restaurants opened on Kearny.[98]

In the late 1960s and 1970s, the city began trying to evict the largely Filipino residents of the International Hotel, at the corner of Kearny and Jackson, in the name of redevelopment. The disruption to the community made it hard for local restaurants to stay in business, and by 1982 only a few Filipino restaurants were left. One, the Mabuhay Gardens, evolved from a restaurant featuring upscale Philippine cuisine into a nightclub, and then into a major punk rock venue. Yet Filipino restaurants were beginning to appear in other parts of San Francisco. In 1983 Virginia Alido opened Alido's in the Mission district after successfully running a small shop selling Philippine ingredients and prepared foods. Alido's was a turo-turo ("point-point") restaurant, where customers point out what they want and then carry it to their tables. Alido soon opened several more local restaurants. In 1994 Ongpin Restaurant opened in South San Francisco; in 1998 the Filipino fast-food chain Jollibee opened its first American restaurant in Daly City, bringing a Filipino version of American-style food to the United States. In the first few months after it opened, customers often had to wait an hour to be served. One dessert treat that both fast-food and family-run Filipino restaurants offer is *halo-halo* (mix-mix), a cheerful mix of shaved ice, tropical-flavored ice cream, and flavored gelatins along with small bits of coconut, sweet potato, plantains, garbanzo beans, corn, or jackfruit. Halo-halo is now so popular in the San Francisco Bay Area that Mitchell's Ice Cream also offers this treat.[99]

Counterculture

Bohemians

In his book *Bohemian San Francisco*, Clarence E. Edwords traced the counterculture back to the early Gold Rush days:

> Cooks [arrived] to prepare delectable dishes for those who had passed the flapjack stage, and desired the good things of life to repay them for the hardships, privations and dearth of women's companionship. . . . Here began the Bohemian spirit that has marked the city for its own to the present day.[100]

From its beginnings in 1863, the Cliff House was a late-night haunt of the Bohemians. San Franciscans stayed up all night and rode over in the very early morning, "when the air is blowing free and fresh from the sea," for a breakfast of broiled turkey and ham, with corn fritters, watching through the western windows "the waves cresting with amber under the magic touch of the easterly sun."[101]

Other favorite late-nineteenth-century Bohemian spots were Babs' Epicurean Restaurant, where one could eat off a coffin to the light of green candles in skulls; Campi's, where a "primitive style of serving" complemented the excellent food; Coppa's, where "walls and ceiling were decorated with the grotesque fancies of its artist frequenters"; and Sanguinetti's, where people went to be "unaffected and natural," until slumming tourists started showing up. Bohemians were also drawn to the cheapest restaurants, to Chinese restaurants, taverns, and coffee shops, especially if they were open late and let people sit for hours over a glass of beer or a cup of coffee.[102]

By the 1870s a group of journalists, artists, actors, and musicians had formed an official Bohemian club, located in two rooms on the corner of

Sacramento and Webb. Soon they were fighting over whether to include people friendly to the arts who were not themselves artistic. The "Ultra Bohemians" said they should stick with "beer and a crust, a garret and sawdust," rather than sell themselves out for "champagne and pâté, marble halls and velvet carpets with the Philistines"—but most of the Bohemians disagreed and welcomed people with enough funds to support and expand the club.[103]

Around the time the original Bohemians sold out for champagne and pâté, San Franciscans began to focus on a darker side of the counterculture—the lure of sex, drugs, gambling, and alcohol. These vices were not new to the city, but residents were newly aware that their city was getting quite a reputation. People traveled from across the United States and around the world to visit San Francisco's "Barbary Coast." Some locals condemned such tourism, but others emphasized the potential opportunities. Pacific Street, the heart of the Barbary Coast, became known as "Terrific Street." Both tourists and locals frequented the groggeries, deadfalls, melodeons, and other tawdry attractions on Terrific Street—drinking wine, beer, or liquor, and hoping to get more than a smile from someone attractive and vulnerable that evening.[104]

Rebuilding after the 1906 earthquake, the city's boosters worked hard to bring back the tourists and were willing to tolerate some suggestive references to decadent, countercultural fun. The *Chronicle* printed a supposed note from a visiting "club woman," seeking to tour "the Bohemian restaurants and cafes that a lady can see with propriety . . . the refined part of the Barbary Coast."[105] Edwords's book, *Bohemian San Francisco*, shared that approach. Publishing in 1914, just before the crowds arrived for the Panama-Pacific International Exposition, Edwords promised a decadent tour of the city. He mostly provided restaurant reviews rather than a guide to illicit pleasures; indeed, he teased the reader by noting that the best fun was to be had in private parties, "in the rooms and apartments of our Bohemian friends." He provided some suggested recipes for these simple potluck meals, such as a ten-cent dinner of tagliarini pasta, or the only slightly more costly dinner of a salad with "Spanish" eggs:

> **Spanish Eggs:** Empty a can of tomatoes in a frying pan; thicken with bread and add two or three small green peppers and an onion sliced fine. Add a little butter and salt to taste. Let this simmer gently and then carefully break on top the number of eggs desired. Dip the simmering tomato mixture over the eggs until they are cooked.[106]

Bohemian women also saved money by entertaining at home. One wrote about her frugal pleasures:

The simplest lunch or tea in a fashionable restaurant is an extravagance. . . . A cozy cup of tea served in your own room will answer all purposes and you not only save money, but you win friends, when you admit them within the portals of your home, even when that home consists of only one room in a boarding house. . . . I bought the best of tea, and served knowing little sandwiches—or delicious little tea cakes which I bought for a penny apiece at the corner shop.[107]

She first invested in a teakettle and electric stove, and later acquired a chafing dish, a percolator, and a "fireless cooker," much like a modern-day Crock-Pot. When she had men over, however, she had to be circumspect to avoid gossip. One rainy evening she welcomed a married couple into her home for noodle soup, a fruit salad, and cold meat with sliced olives and chopped pepper. That simple meal cemented a friendship, and the couple understood their protective role: "[They] make possible many Bohemian feasts in which I could not indulge without such chaperonage."

Gay San Francisco
Unchaperoned sexual licentiousness of all kinds has long been a part of San Francisco's reputation. For gay men and women across America, San Francisco provided a special place where they could be affectionate in public with the ones they loved, even if only during a short stay. Particularly once Prohibition ended in 1933, the city emerged as a place for gay tourists to visit, where they could explore a diversity of gay cultures and communities, eating and drinking alongside other gay people in nightclubs, bars, taverns, and later in gay-owned restaurants and cafés as well. World War II increased the visibility of gay culture, as military personnel arrived in huge numbers. Away from their hometowns, many were eager to experiment. The entertainment and hospitality industries quickly saw the advantage of appealing to these visitors, just as they had played up San Francisco's Bohemianism for an earlier generation of countercultural tourists.[108]

Many gay people were not content with an occasional taste of San Francisco. After the war, soldiers and sailors sometimes stayed in the city to avoid returning to a small-town, secretive life; others began to join them from across the country, seeking to live more openly. Lesbians lived in many neighborhoods across San Francisco, while gay men began building a new kind of community in the Castro.

The Black Cat, a restaurant on Montgomery Street, had Bohemian flair and began attracting a gay clientele in the 1940s. Evenings at the Black Cat featured José Sarria, an early gay activist and drag performer. Customers also came in for the thirty-cent "Catburger," as well as the enchiladas and chili con carne at only a quarter each. On Sunday mornings people came in to recover and rehydrate after a long night of excess, breakfasting for a dollar on pancakes, ham and eggs, toast, tomato juice, and coffee.[109]

In 1948 a gay woman named Tommy Vasu opened a bar called the 299, at Broadway and Front; and then in the 1950s, she opened a restaurant called Tommy's Place, on Broadway and Columbus. Over on Grant, two other restaurants, Miss Smith's Tea Room and The Copper Lantern, also had reputations for attracting a lesbian crowd. One of San Francisco's most visible early gay entrepreneurs was Charlie Marsalli, who ran a gay-friendly restaurant at Mason and Geary during most of the 1950s. In 1958 you could dine there for $2.45, with a choice of steak, prime rib, chicken, or fish. Marsalli called it "San Francisco's Finest Dinner." He also was an owner of 356 Taylor, a gay bar on the next block over.[110]

In the 1960s the wave of refugees from more conservative parts of America became a flood. People who did not fit well with middle-America's standards for gender and sexuality came in droves to San Francisco. The police began to take a harder line on bars and cafés welcoming a visibly gay clientele. In August 1961 they arrested more than one hundred people at the Tay-Bush, an all-night hamburger and beer joint at Bush and Taylor in the Tenderloin. In response, gay activist José Sarria made a serious bid for election to the San Francisco Board of Supervisors that fall, as some parts of the gay community began to piece together a campaign against police harassment. In 1962 a number of gay bar owners formed the Tavern Guild to promote their common interests against official antigay policies and attacks. The raids continued, however, and effeminate or transvestite men found themselves excluded from bars and restaurants on the grounds that they might motivate a police raid.

Some of these people began to congregate at the Tenderloin location of Compton's Cafeteria on Taylor Street. Compton's was a cheap all-night eatery where one could stay warm over a twenty-five-cent bowl of oatmeal or a five-cent cup of coffee. In 1966 the cafeteria put a new service charge of twenty-five cents on all orders; many perceived the fee as a move to push out poor customers who had nowhere else to go. That same year, ministers at Glide Memorial Church helped start Vanguard, an organization for troubled gay and transgender youth. Vanguard's members soon began picketing Compton's, challenging the new fee. Tensions were running high, and one hot summer night a riot started at the cafeteria. According to one version, the riot began when a drag queen threw her coffee in the face of a policeman who was there to harass and remove undesirable customers. Other customers began throwing sugar shakers through the plate glass windows, and the police arrested as many of the drag queens and hustlers as they could in the chaos.[111]

In 1977 Glide Memorial Church was still involved in supporting alternative sexualities, including helping to found the first gay synagogue in San Francisco, Congregation Sha'ar Zahav. Food provided the Jewish gay community with a powerful way to show their fellowship and shared ideals. The idea grew out of a Jewish club called Achvah, "a group of gastronomic Jews who related to their Jewishness by eating chopped liver a few times a year." Once they decided to form a synagogue, food continued to play an important symbolic role. At their first Hanukkah celebration, for instance: "the men made the latkes and the women didn't. In 1977 this was a *big deal*." As they set to the task of rewriting the prayers to suit their social ideals, the leaders sometimes amused themselves crafting lighthearted parodies instead:

> Bless you, She, our almighty but non-aggressive Concept, who . . . suggested to us the eating of vegetable proteins . . . [and] creates the crust of the quiche.[112]

Beats and Hippies
The Beats and the hippies were also marginalized in San Francisco. In the late 1940s Jack Kerouac traveled across the country with various friends but very little money and famously wrote about his memories of being hungry in the city by the Bay:

> I smelled all the food of San Francisco. There were seafood places out there where the buns were hot, and the baskets were good enough to eat too; where the menus themselves were soft with foody esculence as if dipped in hot broths and roasted dry and good enough to eat too. Just show me the bluefish spangle on a seafood menu and I'd eat it; let me smell the drawn butter and lobster claws. There were places where they specialized in thick red roast beef au jus, or roast chicken basted in wine. There were places where hamburgs sizzled on grills and the coffee was only a nickel. And oh, that pan-fried chow mein flavored air that blew into my room from Chinatown, vying with the spaghetti sauces of North Beach, the soft-shell crab of Fisherman's Wharf—nay, the ribs of Fillmore turning on spits! Throw in the Market Street chili beans, redhot, and french-fried potatoes of the Embarcadero wino night, and steamed clams from Sausalito across the bay, and that's my ah-dream of San Francisco. Add fog, hunger-making fog, and the throb of neons in the soft night, the clack of high-heeled beauties, white doves in a Chinese grocery window.[113]

His stories of actually eating in San Francisco—at Alfred's Steakhouse, for example, or the meals of brains and eggs or lamb curry cooked by his friend's girlfriend in a little shack near Sausalito—fade away in comparison with his overwhelming experience of being too poor to buy any of the dishes producing those delicious smells. Kerouac certainly knew the power of a sensory experience: When Allen Ginsberg performed his poem "Howl" at a North Beach art gallery in 1955, Kerouac took up a collection and bought several gallon jugs of cheap California red wine to create a more ecstatic reaction to Ginsberg's performance. Other Beats also sought out the cheapest of cheap food and spent any spare money on drugs or alcohol. The Co-existence Bagel Shop on Grant welcomed poets and their groupies with servings of potato or macaroni salad for a quarter, as well as free-flowing beer.

By the mid-1960s, hippies were becoming a part of countercultural San Francisco. In 1963 a coffeehouse called the Blue Unicorn opened on Hayes Street in the Haight-Ashbury district, advertising the lowest prices in town as well as "food, books, music and art." Soon the Legalize Marijuana movement and the Sexual Freedom League were holding meetings there, and everyone knew you could wash dishes for meals if you had no money. Hippies with a little more money to satisfy their cravings could go to Quasar's for ice cream or Tracy's for doughnuts. In 1966 Jerry Sealund came to the Haight and opened an organic health food store, Far-Fetched Foods (also known as Blind Jerry's), and Fred Rohe opened New Age Natural Foods, also called the Hip Health Food Store. These two were among the first stores marketing their food as natural and hip, rather than as old-fashioned health food.[114]

As more and more people without resources arrived to join the hippie movement, a radical community-action group calling themselves the Diggers began leafleting on Haight Street:

Free Food Good Hot Stew
Ripe Tomatoes Fresh Fruit
Bring a Bowl and Spoon to
The Panhandle at Ashbury Street
4 pm   4 pm   4 pm   4 pm
Free Food   Everyday   Free Food
It's Free Because it's Yours!
    —the diggers[115]

As promised, the Diggers were there in Golden Gate Park every day, passing out free food—not as charity but as a communal action. This took enormous effort, collecting food scraps from grocery stores, from the Ukrainian Bakery, and from the city's produce market. Men were involved at the beginning, but before long Digger women were the ones cooking up big vats of stew after procuring ingredients from the market: "The Italians who controlled the market simply would not give free food to able-bodied men and consequently the women became our conduit to this basic necessity."[116] Digger and poet Lenore Kandel told an interviewer: "I'm a writer, but I'm a woman. . . . I spend a lot of the time in the kitchen, I feed a lot of people all the time. . . . I don't see why I can't do it all."[117]

In 1967 a Hare Krishna Temple on Frederick Street began giving out free vegetarian dishes, called *prasadam*, or offerings to Krishna. Others also began distributing brown rice and vegetables to the growing crowds of hungry young people. San Francisco had had a wave of vegetarianism at the turn of the twentieth century, with two vegetarian cookbooks and a vegetarian cafe on Market Street. But now the interest in natural foods and vegetarianism grew quickly. Within a few years, the Veritable Vegetable collective had started, and Vegi-Food on Clement Street opened—the first strictly vegetarian Chinese restaurant in San Francisco:

> Adhering to strict Buddhist tradition, the cooks don't use eggs, garlic, onions or MSG in the food preparation, and no smoking or alcohol is permitted . . . The house specialty is the award-winning Fried Walnuts with Sweet and Sour Sauce ($4.50) . . . fresh California walnuts dipped in a light batter and deep-fried in cottonseed oil. The sauce is made with fresh orange and lemon juice, chunky tomatoes and pineapple, and absolutely no food coloring.[118]

Vegetarianism was one way of expressing concern for community beyond the Haight. In 1974 Bonnie Ora Sherk and Jack Wickert created an intentional community known as "The Farm," located where Army Street (now Cesar Chavez Street) intersects the US 101 freeway. The Farm was intended to provide a multicultural space where people could "experience plants, animals, and each other in an enriched series of environments within a context of art." By the freeway, they grew fruit, vegetables, and corn; in the kitchen people cooked together and shared food.[119]

In the 1970s people in the counterculture created different kinds of food systems to suit their different needs. In the East Bay, restaurant collectives and radical bakeries (including the Uprisings Baking Collective and Your Black Muslim Bakery) combined politics with food production. The Black Panther Party began serving free breakfasts to Oakland school children, then brought the program to other cities; they successfully pressured the federal government to expand its free lunch program to address breakfast and summertime needs as well. People organized "Food Conspiracies"—clubs for bulk purchasing and for sharing the effort of shopping for organic produce and whole grains. New enterprises such as Seeds of Life (on 24th Street), the Noe

Valley Community Store, and the Rainbow Grocery (on 16th Street) were part of the People's Food System, a network of small community food stores. The Haight Community Store and the Good Life Grocery (on Potrero Hill) followed soon after. These stores had a mission to provide healthy foods, which left them ambivalent about satisfying their neighborhood customers' fondness for white flour, white rice, and canned goods.[120] Many people in the Bay Area, including those on the forefront of the nascent California Cuisine movement, were seeing new connections between their immediate food choices and larger political and economic structures.

Everybody who came to San Francisco had their own reasons, whether seeking greater economic opportunities or the freedom to live the life they wanted. And though their cultures differed, generations of newcomers enjoyed living in a city with so much culinary diversity, where tasty local produce graced the table in ever-changing combinations. All of these many ethnic and cultural groups ended up contributing in different ways to the California Cuisine movement, whether through their imported ingredients, farming expertise, culinary aesthetics, political activism, or a combination of those approaches. Thus San Franciscans of every background feel, rightly, that California Cuisine belongs to them.

CHAPTER FOUR

⁓

Food Markets and Retailing

From the start San Francisco was a market city. It has markets yet, but there is no comparison between them and the old altars from whence arose the incense to the god of gastronomy.[1]

So wrote Walter Thompson, columnist for the *Chronicle*, looking back at the heyday of the city's market culture in the late nineteenth century. He waxed nostalgic for the Washington Market, whose grand arches opened onto Washington Street with other entrances on Merchant Street and Sansome:

Huge sides of beef hung in solemn stateliness from its many-hooked racks— flocks of sheep minus their clothes hung in pink profusion between them. . . . Festoons of plump sausages swung from hook to hook with all the abandon of a popcorn chain on a Christmas tree, and pear-shaped hams by the gross.

Then, too, there were the rolls (not squares) of golden butter, piled in pyra-midal form after the fashion of cannon balls at an arsenal . . . bushels of eggs heaped up like cobblestones; cheeses of all ages and dimensions; fruit in such variety that it seemed as if the Hesperidian gardens had come to town to stay . . . barrels of pickles, tubs of nuts.

Those barrels and tubs provided cover for children playing hide-and-seek while their parents shopped for the week's groceries and then carefully orga-nized their market baskets for the walk home. Thompson also wrote affec-tionately about the stall men, gossiping as they sliced meat, made change, or wrapped eggs in brown paper. By the time of Thompson's article in 1916, the venerable Washington Market had been replaced by the California Market, and other markets had also sprung up, such as the Bay City Market, the Sutter Market, and various namesakes of the Washington Market (first a block away, and later on Market Street).

As San Francisco began to take shape during the Gold Rush, wholesale and retail food provisions played a prominent role in the city's thriving economy. In the beginning, the city's markets only supplied a limited assortment of goods, particularly during the winter months when fresh produce was hard to find. One journalist visited Santa Cruz in 1853 and reported that gardeners in that region promised to end San Francisco's seasonal dearth of fresh vegetables; warmer winter weather in that coastal area meant "peas and other garden vegetables can be produced every month in the year."[2] He went on to imagine a future in which San Francisco would wake up to the arrival of a steamer from Santa Cruz, "laden with peas, beans, squashes, turnips, radishes, lettuce, cauliflower, strawberries, raspberries, gooseberries, currants, cherries, plums, apples, peaches, pears, quinces, melons, and not least of all, good fat geese, turkeys, ducks, chickens, eggs, and many other things." In the meantime, he found it embarrassing that even the best San Francisco restaurants rarely served poultry, and eggs were too expensive for ordinary San Franciscans to enjoy for breakfast—unless they used seabird eggs from the Farallon Islands. Something had to change.

Fifteen months later, newspapers reported that the situation was improving:

> Strangers visiting San Francisco and passing through the markets, cannot fail to be astonished at the vast variety of meats, game, fish, vegetables and fruits which are here displayed. Luxuries from every clime abound in the greatest profusion. . . . No people in the world live more luxuriously than do the inhabitants of San Francisco.[3]

After that puffery, however, the article admitted that, when compared with the East Coast, San Francisco markets offered inferior cuts of beef, veal, pork, and mutton, and chicken was still very expensive. Many Californians were hard at work trying to solve those problems.

At least fish and game were of fine quality and easy to find in the markets. One could buy bear, venison, rabbits, squirrels, quails, pigeons, snipe, curlew, and plover, as well as geese and ducks "of every variety and in the greatest profusion." Salmon, skate, sturgeon, codfish, rockfish, smelt, herrings, sardines, flounder, sole, perch, crawfish, lobsters, shrimps, and crabs were all plentiful, as was blackfish from Southern California and mackerel from San Diego. San Francisco markets always had oysters, but they proved small and disappointing compared with East Coast oysters.

Fruits and vegetables were the highlight of the marketplace: By combining produce from the region's hills and valleys with imports from Oregon, Southern California, and overseas, San Francisco succeeded in presenting incredible diversity all through the year. Eggs, butter, and luscious cheeses were also available by 1854, making up for the difficult early years. People also still bought the strangely colorful seabird eggs, having decided they were especially nutritious. (About five hundred thousand murre eggs were gathered from the Farallon Islands in 1854, and millions more over the next few decades. The murres were on the verge of disappearing. Fortunately, demand for their eggs collapsed in the 1880s as consumers began disparaging the eggs' "rank fishy taste.")[4]

California's bountiful offerings by the mid-1850s held out a promise of even better selection in the future: "California will be able to boast of markets as extensive, as substantially and neatly constructed, and as well arranged and abundantly supplied as any markets in the oldest cities of the Atlantic States."[5] But they were not quite there yet. Part of the problem was that in the 1850s, few San Franciscans had facilities at home to do much cooking. A decade later, a rise in home cooking brought about those abundant offerings Thompson so lovingly recalled in 1916.

Wholesale Markets

By the mid-1850s, agriculture encircled the growing city, and getting produce from farm to middle men required some centralization. For the most part, this took place at the Central Wharf (now Commercial Street), stretching out two thousand feet from Sansome Street to deep water. Restaurant owners and retailers were used to coming there to buy food shipped long distances to San Francisco, and farmers around the bay found it easy to ferry their fresh produce to the Central Wharf.

By the 1870s, Sansome Street from Clay to Washington had become encumbered with the fruit and vegetable trade. Many of the people involved in the trade were Italians; in 1874 they formed the Gardeners' and Ranchers' Association and promptly arranged for the construction of an up-to-date produce market on Davis Street between Pacific and Jackson. When it opened in 1876, a cheerful arch announced the name—Colombo Market—while smaller arches with brightly painted numbers framed each stall, looking out at the street.

Colombo got moving around 3 a.m. in the summer, as ships unloaded and wagons trundled in, bringing produce from market gardens run by northern Italians. Cultivators, exhausted from loading the cart, often dozed during the ride, confident that their horses would get them safely to market. Once at Colombo, farmers carried themselves with "the airs of a millionaire," even though most of them leased their land (turning over about 45 percent of their proceeds to the owner).[6] But they were justifiably proud of their fine produce and socialized happily over a glass of wine or a cup of black coffee at a nearby saloon. Almost everyone in the wholesale market spoke Italian, though various dialects revealed the differing origins of the farmers, the vendors (commission merchants), and the small-scale vegetable peddlers who frequented the Colombo site.

Along a central aisle, growers unloaded crates of cabbages, cauliflowers, lettuce, leeks, celery, eggplants, beets, radishes, turnips, carrots, garlic, and tomatoes, arriving from Sunnyside, Oceanview, Colma, and other gardens southwest of downtown. By 1903 the list of vegetables would also include spinach, onions, red cabbage, stout French carrots, chicory, artichokes, parsley, sage, and thyme. (Vegetables such as potatoes, squash, green corn, peas, and beans, as well as stone fruit, tended to come to San Francisco by steamer from across the bay, rather than from the Italian market gardens.) Visitors found themselves charmed by "heaps of cabbages to a height of eight or ten feet . . . alternating with red beets and vegetables of every kind and hue," even if the market's odors were overpowering. Breezes off the bay helped mitigate the predominant smells of cabbage, garlic, old trampled vegetables, and the ubiquitous cigarettes and pipes.[7]

Colombo's vendors sold their produce to the big retail markets such as the Washington Market, to neighborhood grocers, to restaurants, to boardinghouses, and to peddlers with their carts. Those transactions went smoothly for the most part, especially early in the morning when vendors were simply filling large prearranged orders, or else when the customers were themselves Italian. After the major selling was finished around 7 a.m., Chinese buyers came in, bargaining to get leftover bruised or tattered produce for greatly reduced prices. Those buyers in turn distributed vegetables to Chinese grocery stores as well as to "basket men" who sold their wares door-to-door in residential neighborhoods. One journalist criticized both the basket men and their customers:

> It is an extremely convenient system for lazy and shiftless housewives. A few of these basket-men have gardens in the suburbs, where they live and raise their produce; but the great majority purchase the refuse of the city markets, and by heathen arts that must be nameless, impart apparent freshness to the originally wilted and unsavory stuff.[8]

On the contrary, households looked forward to the arrival of these visitors, who often had little surprises in their baskets. Along with the fresh fruit, greens, onions, and potatoes, they might pull out silver smelts or red shrimp, or they might have tiny bags of Chinese candies for the children who awaited them: "How they carried those loaded baskets up our steep hill has always mystified me. They must have been very strong."[9]

Tourists were fascinated by the Colombo Market. On a visit to San Francisco in August 1915, Rudolph J. Walther walked through with friends, admiring the peaches, strawberries, loganberries, grapefruit, melons, and a large assortment of garden vegetables, including "quite a few that were strange to us."[10] He was amazed that one could buy a crate of plums for just fifty cents, but, wary of damaging them on the way home to Philadelphia, he bought canned California olives instead. One journalist preened about the San Francisco produce:

> When all the rest of the world eats "California fruits" from tins, and considers them a delicacy at that, San Francisco markets have these same fruits—the peaches, apricots, plums and cherries—fresh from the trees of the Santa Clara valley and the orchards of the interior valleys.[11]

Other produce markets sprang up nearby. Farmers who were not part of the Italian network at the Colombo Market paid fees to have their produce carted from the docks or rail center to these markets, where commission merchants bought their goods for resale. Truck farmers parked willy-nilly and sold small wholesale lots directly off the back of their wagons, while nearby merchants complained about garbage obstructing the streets. After the 1906 disaster, the city developed the district, while maintaining it as the central distribution point for San Francisco produce.[12] In the 1960s, after decades of corporate lobbying for access to this prime real estate, the Golden Gateway and Embarcadero Redevelopment Projects pushed out the produce markets. In 1963 the San Francisco Wholesale Produce Market opened at Jerrold Avenue and Toland Street, conveniently near the intersection of Highways 101 and 280 in the Bayview neighborhood, where it remains to this day.

Wholesale Grocers and Importers

For bulk items and processed food products, retailers sought out large whole-sale dealers and importers. Many of these wholesalers were Jewish: the Haas Brothers, for instance, as well as the Castle Brothers, the founders of Ehrman

& Co., and two of the founders of Goldberg, Bowen & Co, Jacob Goldberg and Louis Lebenbaum. (Goldberg, Bowen & Co. also sold retail; see next section.) Other prominent wholesalers included Bigley Bros., William Cluff & Co., and Root & Sanderson. Among the more visible Chinese wholesalers in the late nineteenth century were Tuck Chong and Chy Lung & Co., as well as the Eastern import-export house Karanjia Co., importing rice, tea, peanut and sesame oils, star anise, cinnamon, and various Chinese preserves. Jann Mon Fong, who emigrated to the United States in 1931, later became the first Chinese member of San Francisco's Wholesale Grocers' Association.[13]

Success in these businesses depended on luck and on having a good sense of timing. In the early years, wholesalers made their profits largely from reselling coffee, tea, sugar, and spices. But over time, many shifted into other areas of food commerce. The three Castle brothers, for instance, came over from England during the Gold Rush and parlayed their international connections into a profitable wholesale grocery business. Over time, however, the company shifted into a fruit drying, canning, and packing company, preparing California seedless raisins and other fruit for export around the world, as well as producing honey for domestic and international consumption. Similarly, Wellman, Peck & Co. started in a small building on Kearny Street, importing groceries from Mexico, Central America, and British Columbia but also as far as Polynesia. They specialized in bulk goods such as farina, tapioca, buckwheat flour, pancake flour, and "holgrane" flour. Then, in the twentieth century the firm began marketing and exporting its own products with the "Flavor Famous" tagline: marmalade, peanut butter, coffee, diced beets, jellies and preserves, canned peaches, and canned "Snow-Flake Salmon." F. Daneri & Co. specialized in importing groceries from Europe, such as French and Italian wines, Italian olive oil, macaroni, cheese, and fruits, as well as Central American sugar and coffee.

J. A. Folger & Co. was a prominent coffee wholesaler from the 1860s when Folger took over his former employer's coffee roasting company. Born in Nantucket, Folger had family shipping ties and soon made a name for himself combining coffee imports and bulk-roasting coffee for retailers. More and more, Folger's also sold coffee ready for the pot: roasted, ground, and packaged in "aroma-tight tins."[14] Folger's Golden Gate Coffee, labeled with a picture of a ship in San Francisco Bay, was the company's most expensive blend. Packaging became key to selling coffee. Hills Bros. Coffee won market share from Folger's with a parchment lining to protect coffee from the "many disagreeable odors of a grocery store." Later its vacuum-sealed coffee cans similarly reassured retail customers that the natural coffee flavor would last. During World War I, Central American coffee growers needed a closer market for their crop than war-torn Europe; San Francisco wholesalers were delighted to step up their imports of the flavorful coffee and maintained those relationships after the war.[15]

Retail Markets

San Francisco had numerous retail markets over the years, covered alleys with stalls where vendors sold meat, dairy, fish, and produce. Customers could also purchase prepared foods to eat there or carry away; they could enjoy a beer and the opportunity to people watch. Among the most prominent of these markets were the Washington Market (on Washington Street), the California Market (on Pine, between Montgomery and Kearny), the Bay City Market (on Market Street near City Hall), and the Sutter Street Market.

During the 1882 Christmas season, the California Market boasted such a variety of produce as would astound "residents of the snowy East" and "give zest to the Christmas feast"—new asparagus, new peas, new potatoes, fresh rhubarb, fresh strawberries, fresh mushrooms, along with oranges, apples, pears, bananas, and more obscure fruits such as barberries. Oranges were popular at Christmas, but in December their taste was not yet at its peak; people bought them as seasonal decorations to give their stores, restaurants, or homes some holiday flair. The market also offered magnificent turkeys and geese, as well as many kinds of duck: canvasbacks, mallards, teals, and widgeons. The oyster dealers at the market in that season sold canned "fresh-frozen" Eastern oysters, as well as fresh local oysters from Dumbarton, "a locality which is said to produce the most delicious bivalves on the Pacific coast." The market was conveniently located "on the homeward route of brokers and solid businessmen," who would thus find it easy to make their holiday purchases.[16]

The California Market was also known for some of its characters, such as Michael B. Moraghan, an Irish immigrant known as the Oyster King. Moraghan saw that one solution to the exhaustion of California's local oysters lay in importing Eastern oysters and cultivating them in the Bay Area's mild climate. The hard part was keeping oyster beds level on the bumpy bottom of the bay. His company had hundreds of acres of oyster beds on the tide flats at Millbrae, and kept fifty men hard at work leveling the beds at low tide and fighting off oyster pirates including the young Jack London. Moraghan's main competition, the Morgan oyster company, had even more beds, until urban runoff and environmental changes in the 1910s made the bay unsuitable for oyster farming.[17]

Moraghan, a genial man and friend to San Francisco's Bohemians, ran a seafood restaurant and shop at the California Market. Customers were drawn to his stall by his beautiful oysters. In 1922 Moraghan's was sold to a Croatian immigrant named Sam Zenovich, and over time became Sam's Grill and Seafood Restaurant. The restaurant maintained its reputation for excellent seafood and its home at the California Market until 1946, when it moved to Bush Street.

J. H. McMenomy was another California Market character. A dealer in stall-fed beef, McMenomy came to work every day in a silk top hat and charmed the press:

> J. H. McMenomy says, if you think there is no good beef in the country, call at his stalls, 8 and 9 California Market, and see the best meat you ever saw. He will not show you a lot of trash hanging up to make a big show, but he will show you quality.[18]

When the market first opened in 1867, McMenomy was just one of fifteen butchers on-site; he then spent every day for fifty-seven straight years manning his butcher stall and got to know almost everyone in San Francisco. After his retirement, he reminisced that in those early days one could order lunch at one of the market's restaurants and receive not just a steak and half a loaf of French bread but "in addition shrimps or crabs' legs, or perhaps pigs' feet—all at a cost of twenty-five cents!"[19]

Fisherman's Wharf

San Francisco's commercial wharves stretched from Meiggs Wharf near Fort Mason around to Hunters Point to the southeast. Shipbuilders, dockworkers, fishermen, and fishmongers of many ethnic backgrounds made the wharves a space for hard work but also some small pleasures: smoking in the sun, spitting on the ground, drinking, and watching cockfights. Over time, other San Franciscans began strolling to the wharves near North Beach on fine Sundays for swimming, boating, or fishing. These crowds attracted vendors selling fresh bait for Sunday fishers, fresh cracked crab for hungry strollers, and foaming lager beer for the thirsty. Starting in the late 1850s, Abe Warner attracted visitors with a menagerie of monkeys, baboons, bears, and parrots at his Cobweb Palace, located on Francisco Street near the waterfront; customers then slaked their thirst with sherry cobblers, ale flips, and rum toddies and enjoyed an excellent "Sabbath chowder" made from crab or clams, depending on the season. From the 1860s, fishmongers boiled crabs in huge pots on

Figure 4.1.   Crab stand at Fisherman's Wharf. Courtesy of the Union Pacific Railroad Museum.

the wharf; they began serving fresh crab alfresco: ten cents per crab, with the crab-cracking mallet furnished for free. A nickel bought a glass of beer, along with piles of crabs' legs to munch for no extra charge.[20]

In the 1890s the new Powell Street Wharf drew women with their children down to the waterfront:

> It is a beautiful place and affords a splendid view of the bay in every direction. But the reason the ladies like it is because they can fish there without attracting the attention of a crowd of wharfrats. It is now no uncommon sight to see a well-dressed lady hauling up a big crab net.[21]

Crab was king on the wharves, but innumerable kinds of seafood were caught and sold there as well. In 1904 the tomcod came in, swarming around the newly built Fisherman's Wharf at the base of Taylor Street. No one had ever seen fishing like this. Amateurs were tossing in lines and coming home with baskets full of fish:

> Men and women were there from every quarter of the city. . . . Girls from the art schools, who came to sketch, remained to fish. . . . Girls who had come to

write up the local color squeezed with gratitude between dirty little boys of the joyous water front. . . . It was time for supper, but the mother was hauling up tom cod at the rate of two a minute and the excitement had gone to her head.[22]

Everyone was there, sharing in the thrill of the tomcod "rush." From that point on, Fisherman's Wharf became central to San Franciscans' view of their shared city.

In the 1920s some of the Wharf's seafood shacks began developing into real restaurants, with official permits from the city. They served Crab Louis and cioppino, as well as cracked crab and chowders. Among the earliest restaurants were Alioto's, Castagnola's, and Sabella's. The long-established DiMaggio fishing family opened DiMaggio's Grotto on the wharf in 1937, soon after Joe DiMaggio's major league debut with the New York Yankees. The new restaurants and tourist trade did not, however, replace the older uses of the Wharf as a base for fishing and a marketplace for fish and shellfish.

Commercial and ethnic disputes over fishing rights date back to the Gold Rush era. At first, the Chinese were criticized for overfishing, or for catching too many fish in their fine shrimp nets. By the turn of the twentieth century, the same accusations bounced back and forth between Genoese and Sicilian fishermen. Northern Italians had dominated the docks throughout much of the nineteenth century, but immigrants from southern Italy began challenging their control. By the 1910s local Italian fishermen had divided,

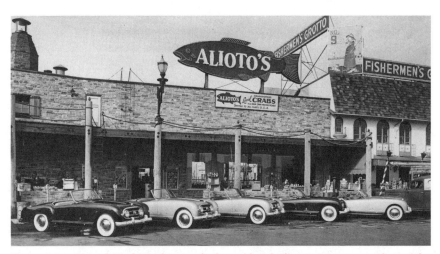

Figure 4.2. Several Nash-Healeys parked outside of Alioto's restaurant. Photo John Black and Associates. Courtesy of the San Francisco History Center, San Francisco Public Library.

Figure 4.3. Castagnola restaurant, Fisherman's Wharf. Courtesy of the San Francisco History Center, San Francisco Public Library.

by their region of origin back in Italy, into competing organizations. Among the larger organizations were the Crab Fishermen's Protective Association, the Western California Fish Company, Paladini Brothers, the Borzone Fish Company, the International Fish Company, and the F. Alioto Fish Company. Each focused on keeping prices high for their stock in trade, which occasionally meant intimidating competitors. Some fishermen stayed out in their boats for days at a time, curling up to protect themselves against the cold and eating cioppino—the catch of the day, cooked into a hearty stew with lard, onions, garlic, chilis, and tomatoes. With new fishing technologies, capital and organization—rather than physical strength and sheer grit—began to determine success. Large companies controlled much of the local business and, at times, colluded to keep prices high. During World War II, many Italians were excluded from the city's wharves as possible enemy sympathizers, creating hardship for those families.[23]

San Franciscans have long complained that tourists were turning the Wharf into an imitation of its former self. In 1961 a guidebook noted: "The big boom started in the 1940s and is still continuing. Now the wharf is lined on every side with gaudy restaurants, gift shops, and sideshows with assorted sea monsters."[24] In 1964 the historic Ghirardelli chocolate factory near Fisherman's Wharf was repurposed into trendy Ghirardelli Square, one of the first such redevelopment projects in the United States. Nearby, the old Del Monte fruit cannery was soon likewise transformed into a vibrant spot for retail and restaurants. Local fish canneries had left San Francisco for Monterey by 1900, but near the wharves the Del Monte Cannery processed peaches and other fruit from 1907 to 1937. In 1966 *Chronicle* columnist Herb Caen declared the new Cannery marketplace a "peachy" idea.[25] With their early twentieth-century atmosphere, Ghirardelli Square and the Cannery helped revitalize Fisherman's Wharf into a popular destination for both tourists and locals.

The Meat Trade

Meat had been a big component of Yerba Buena cuisine, and already in 1847 the town had at least seven butchers. With the Gold Rush, many more men joined them, although most did not have much training in the trade. Brannan Street was known as Butchertown for more than a decade, until growing numbers of local residents complained about the smell and the "many accidents and hairbreadth escapes . . . when droves of long-horned, wild steers were driven through the crowded thoroughfares of the city." In 1871 Butchertown moved to the Bayview district; the cattle arrived from the south, and gin shops grew up alongside the new, "thoroughly clean" slaughterhouses. Fist fights and dog fights proliferated, providing butchers with some variation in their daily blood and gore.[26]

In 1874 the California Legislature passed the No-Fence Law, marking a shift in influence from ranchos to agriculture. Cattle trampling crops had been the farmer's problem unless he had extensive fencing around his fields. The new law made ranchers responsible for the actions of their livestock, which in turn made California cattle very expensive to maintain. At the same time, the new railroad system made it affordable to transport cattle long distances: "Nevada thus became practically a great pasture-field, [supplying] San Francisco with meat." Once slaughtered, the "rich, red meat, clouded or smothered with delicious flakes and folds of fat," was purchased by the city's retail butchers and sold as quickly as possible to the consumer, especially during the hottest days when the meat would soon begin to spoil. In 1879 about three hundred cattle, six hundred hogs, and two thousand sheep were slaughtered each day in San Francisco. These vast quantities of meat meant that "our people, however poor, eat their juicy, sweet and healthful meals of meat three times a day, and at a mere nominal expense." A housewife could buy steaks for five cents a pound; pork was even less.[27] The wholesalers' willingness to extend credit to favored retailers led to tight relationships but prevented San Francisco's retail butchers from organizing effectively to defend their interests against the wholesale industry.

Retail butchers put their efforts into impressive displays of meat, wooing customers to the city's markets. Holiday seasons summoned forth exceptional exhibits, such as butcher Mark Strouse's spectacle at the Bay City Market: In December 1889 his stall featured hundreds of cattle, sheep, lambs, and "porkers ranging up to 1270 pounds in weight," as well as two black bears. The bears were apparently a hit; a couple of years later Strouse included three huge bears in his Christmas display: an Alaskan bear, a British Columbian bruin, and a California cinnamon bear. And, in 1895, he displayed a huge grizzly bear, just when the California grizzly was becoming extinct. Butchers mostly sold meat from bears captured as cubs, each one stall-fed until he reached a good weight: "He will eat the same food a hog will eat and about the same quantity; and his flesh tastes very much like pork except for a gamy flavor. . . . His hind-quarters furnish superior hams, and his ribs yield the best of bacon."[28]

In the winter of 1890 Strouse began peddling meat in refrigerated wagons, selling door-to-door to his many customers. More than two hundred San Francisco butchers organized against him, complaining that he was taking all the small cash customers and leaving to the butchers' shops those less attractive customers who relied on credit to buy their weekly meat. The butchers persuaded the Board of Supervisors to put in a new ordinance, requiring peddlers of meat to buy a license of seventy-five dollars per vehicle every quarter. That slowed down Strouse but did not prevent the incursions of refrigerated meat. P. D. Armour, who first came to California during the Gold Rush, expanded his midwestern meatpacking business to San Francisco in 1894. He sold refrigerated and canned meats to the city's consumers despite combined action by the city's wholesalers and retailers against the new Western Meat Company.[29]

Over the next few decades, struggles between union butchers and non-union shops sometimes flared up into violence. Chinese butchers, excluded from the San Francisco Butchers' Union, were often victims of union actions. In 1887 Strouse, as vice president of the union, called for shutting Chinese butchers out of the pork and sausage industries. He decried sausages made of "Chinese-fed pork" and proposed to send a wagon all over the city plastered with the names of every shop that bought meat from Chinese butchers. The union's antagonism toward the Chinese continued over decades, despite language in the bylaws of the American Federation of Labor granting equal rights and privileges to all unionists without distinction of nationality, race, or color.[30]

In San Francisco, a woman could become a butcher if she inherited the business from her husband. Fredericka Junker ran the family butcher's shop at the corner of Oak and Gough after her husband, Charles, died in 1891. She admitted that some of her customers were drawn by the novelty of seeing a woman do a traditionally male job. A few years later, the Alsatian immigrant Frances Daverkosen gained a similar level of fame as "the only woman butcher in San Francisco." She had learned from her second husband how to cut and trim meat for their customers; after he died, the business thrived under her management:

> A gray-haired, mild-faced, benevolent-looking old lady, who beamed through spectacles as she sharpened a twenty-four inch knife. . . . She has hanging about her some of that charm which never deserts a French woman. . . . She is the equal at cutting meat of any man in the city.[31]

Junker and Daverkosen both seemed to take pride in their nonconformist roles, even as they turned back quickly from a press interview to the work at hand.

In the twentieth century, supermarket meat departments increasingly replaced neighborhood butchers. Since the 1970s, however, some consumers have sought out sustainably raised meat and supported the Niman Ranch business model. In recent years, that movement has sparked a revival of small shops where experienced butchers explain the less well-known cuts of meat to San Francisco's home cooks.

Bakeries

San Francisco is now known for sourdough bread, but in the nineteenth century each ethnic group in the city had its own bakeries with unique breads. Already in 1854 the city directory listed sixty-three bakeries, from the American Bakery to Washington Bakery, as well as two named "French Bakery," two named "German Bakery," the Franco-American Bakery, the Hamburger Bakery, the Lafitte Bakery, the Louisiana Bakery, two named "New Orleans Bakery," the Patent Steam Bakery, and Winn's Fountain Head Bakery. Ten years later, those numbers had barely shifted, but by the mid-1870s there were almost three times as many bakers, with names such as Deutsch, Heitmuller and Schneider; Hirschfeld, Kahn and Meyer; McCann, McCloskey and McGrath; Boudin, Guerin, and Verges; and two bakers named Sanchez. Starting in the 1880s bakeries across San Francisco faced a series of strikes as workers fought for higher wages, shorter work days, and a day off once a week. In the 1890s the growing "Pure Food" movement drove public interest into the disturbing sanitary conditions at many bakeries. One inspector reported "the places we have visited are so utterly filthy that I am surprised that the outrageous condition of things has not been exposed before."[32] Despite these concerns, most families in San Francisco continued to rely on bakeries for their daily bread, rather than making their own.

In 1904 the *Chronicle* published a major, illustrated feature on bread, which set the tone for the city's bread marketing during the twentieth century. The article was provocative, titled "San Francisco Uses More Kinds of Bread Than Any Other City in the Country," and subtitled "You Can't Buy a Loaf of Real French or Italian Bread Anywhere Else in the United States."[33] It made the case that San Francisco should be nationally known for its bread:

> According to flour experts, San Francisco bakes and eats more varieties of bread than any other city in the country. . . . There is the American bread, the French bread, the Italian bread, the Mexican bread, the German bread, the Russian bread, the Greek bread, the Swedish bread, and varieties of these.

The most common bread in San Francisco was Irish bread, or, as it was also known, "family loaf." Irish bread was oblong, baked in tins, and had crusts only on the top and bottom of the loaf. Two loaves of this bread cost five cents. The next most common bread was called milk bread, and it was made of milk, flour, water, and salt. The shaped loaf was placed directly on the floor of a brick oven, and a crust formed all around it.

In the 1880s San Francisco's bakers had used only California flour in milk bread—"which made it sweet, as are all California breads." But tastes changed as the city's population changed: "Eastern people coming here liked a very fine texture," so bakers began adding lard and somewhat tougher eastern flour, balanced by extra sugar as well. In San Francisco, the bakers of these Irish breads and milk breads were almost all German. German bakers also made pumpernickel and rye breads, adding caraway seeds for variety. Russian and Polish bakers made heavy, dark ryes, whereas one Swedish specialty bread involved a mix of rye and wheat, rolled out flat, cut into circles, punched with little holes, and seasoned with caraway. Most San Francisco breads were made with yeast, whether the flour was white, whole wheat, graham, or rye. The exceptions were pumpernickel, French, and Italian bread.

Of those, the French sourdough was the one most similar to San Francisco's sourdough bread today and was served in many restaurants, not just French places. In contrast, Italians were the main consumers of the round Italian loaf: "A very dry, sour bread, and, because of its dryness, it keeps very well. . . . Fishermen use it a great deal on their trips."

The article touted the recent introduction of industrial machinery into local bakeries: mixers, kneaders, shapers, and conveyor belts, as well as new patent ovens. Early twentieth-century industrialization hurt many of the city's small bakeries, a common story across the United States. By 1909 the area's two largest bread makers, the California Baking Company and Homestead Bakery, were putting out about two hundred thousand loaves of bread each day, more than all of their competitors combined.[34] Mass-produced bread took over until a new wave of artisanal bakeries emerged in the 1970s, including Il Fornaio and Tassajara Bakery. Acme and Semifreddi's followed in the 1980s, joined by the San Francisco Baking Institute and Pascal Rigo's Boulangerie in the 1990s, and then Tartine, which Mark Bittman called his favorite bakery across the United States. Rather than maximizing shelf life, these bakeries focused on taste and texture, on achieving a crunchier crust and a more flavorful crumb.

Grocery Stores

Corner Groceries

In her wonderful book on women navigating public spaces in turn-of-the-century San Francisco, Jessica Sewell explains the role of the city's corner groceries and delicatessens. In certain neighborhoods, much of the city's bounty came right to one's doorstep, as in the Levy home mentioned earlier. In other parts of town, however, people regularly visited the local grocer and other food purveyors. Apartments had small pantries and iceboxes, so people went to the corner grocery for each day's supplies. Housewives did not handle the goods themselves. Until the 1920s, the grocer or a clerk selected, weighed, and packaged the desired items; often a delivery boy would carry the packages to one's home. Women had their choice of grocery stores in the neighborhood, so successful grocers were the ones who went out of their way to make their store feel safe, warm, and almost domestic. The grocer helped shoppers make good choices and made suggestions for how they might use new products, or the freshest produce. Grocers wanted customers to feel comfortable sending their young children to the store to pick up some forgotten item. Many grocery stores allowed longtime customers to buy on credit, so

one could send children on this errand without worrying about them drop-
ping the money along the way.[35]

A sizable minority of the grocers were women themselves, whether under
her own name (about 7 percent in 1890, about 12 percent in 1911); with her
spouse; or under her deceased husband's name if she outlived him. Almost all
of the women who ran their own grocery stores lived right above or behind
the store, making it a convenient career for a single woman. Even when a
couple worked together as grocers, they usually lived right near the store. The
Retail Grocers' Advocate declared in 1908 that women were often better than
male grocers at answering any cooking questions their customers might have.
Overall, however, the journal asserted that women grocers were best as a
cheerful face in the grocery store and should leave the business side to men.[36]

Around 1890 grocery stores were caught up in a controversy over women
and drinking. Groceries made a sizable profit from selling alcohol to the
neighborhood, whether bottles of whiskey, pails of beer (called "growlers"),
or a glass at a time to be consumed right there at the grocery store. One
observer complained that children picked up beer for their mothers to drink
on the sly:

> "Rushing the growler," "flying the duck," or "chasing the Mickey," as it is vari-
> ously termed, is a time-honored pastime in San Francisco, and Tar Flat is by no
> means the only locality in which processions of little children pass to and fro
> with the beer can. . . . The man who keeps the corner grocery first weighs the
> pitcher or can, and then supplies the beer by weight, at the rate of two-and-a-
> half pounds for ten cents.[37]

Other children asked for a bottle of whiskey; moralizers were scandalized to
think of mothers teaching their daughters to "conceal the bottle in the folds
of their scanty dress." Sometimes grocers cooperated to help housewives
drink in secret, submitting fraudulent bills to the husband where whiskey
showed up as macaroni and beer as soap.[38] Women also went to the grocery
to drink, entering from a side door to avoid neighbors who might have
stopped by for milk, flour, or other more wholesome supplies. Temperance
supporters worried that "a woman may violate the criminal and the social
code by getting as drunk as she pleases," knowing that she was only a few
steps from her own house. In 1893 the San Francisco Board of Supervisors
passed an ordinance against "ladies' entrances" to saloons, but despite some
criticism they left corner groceries alone.[39]

Grocery stores were spread all over the city, in every residential area. From
1890 to 1911, the number of stores barely budged, hovering around twelve

hundred grocers, more than the number of butchers, bakers, and produce dealers put together. They were usually in small storefronts, using all the space available. Unlike dry goods stores with their formal window displays, grocers put tables of ripe fruits and vegetables in front of the store. Without even leaving the sidewalk, passersby were already invited to think of themselves as customers. Grocers paid attention to what the neighborhood wanted, so Chinatown groceries carried not just basic supplies such as flour, sugar, coffee, and tea but also Shantung cabbage, water chestnuts, and lotus seeds, as well as preserved Chinese fruits, vegetables, and fish. Grocers in North Beach sold many kinds of olives, olive oil, holiday cakes, and boxed noodles; Russian grocers focused more on canned beets and caviar. Grocers also often sold prepared foods, ready to eat, either on the spot or for take away.[40]

Goldberg, Bowen & Co.

In the late nineteenth century, many of the well-to-do in San Francisco satisfied their gourmet desires at Goldberg, Bowen & Co. The upscale grocery delivered imported goods and fine wines across the city. The Bowen brothers came to San Francisco in 1859 and first set up shop at the corner of Montgomery and California. A decade later, they moved their grocery store to the new California Market, off Pine Street. There they built their network of connections to suppliers all over the world, until they gained a reputation as San Francisco's preeminent grocery establishment, both wholesale and retail. The grocers advertised their long reach, promoting codfish shipped from Newfoundland ("just arrived per steamer, via Panama, in splendid condition"), St. Louis hominy, Stewart's crushed sugar from New York, French sardines, Siberian salmon bellies, and canned pineapple from Singapore. They also emphasized local foodstuffs: at a horticultural exhibition in 1870, the Bowen Brothers displayed a wide range of Bay Area products including "wines, jams, pickles, dried fruits, starch, vermicelli, honey, and a host of other articles too numerous to mention."[41]

In 1881 they took on new partners; for the next sixty years the upscale store was known as Goldberg, Bowen & Co., or variations of that name. The store's prestige continued to expand. Customers enjoyed their prepared foods as well, and restaurants advertised that they served hors d'oeuvres from Goldberg, Bowen & Co. After a devastating fire in November 1894, which destroyed $250,000 of stock (largely liquors, cigars, and fancy delicacies imported with an eye toward the upcoming holidays), the branch on Sutter Street at Kearny took on the flagship role. The grocery store, then called Goldberg, Bowen and Lebenbaum, sponsored a student poetry competition that same year. The winning poem, by Berkeley sixth grader Hazel A. Brown, presented an adoring view of the grocers' mission:

> Groceries
> In every country, state and clime,
> Groceries are needed all the time.
> Where, the freshest and best of them to find,
> Is ever the wish and study of mankind.
> Blest is the firm who, with the greatest of ease,
> Has found the way the people to please.
> The grocers, Goldberg, Bowen and Lebenbaum,
> Have joined together and found the charm . . .
> The minute you enter the very door
> You see many things you ne'er saw before;
> And whenever a trip through the store is paid,
> You may well think a trip round the world you have made . . .
> The groceries kept are always fresh and good,
> And embrace every known article of food . . .
> So may Heaven bless and keep from harm,
> Our grocers, Goldberg, Bowen & Lebenbaum.[42]

The main store on Sutter Street was destroyed in 1906, and contemporaries noted that loss as a particular blow. Three months after the earthquake, one relatively well-off San Franciscan, Mabel Coxe, wrote to her brother in the Philippines:

> Oh, Charlie, it was a dreadful, dreadful time, and is pretty bad yet, for that matter. . . . At first, that is the first two weeks, there was nothing, absolutely no place where you could get anything. Those who had forethought on the morning of the quake laid in provisions, for they knew the worst was yet to come. Some of the grocerymen took advantage of the state of affairs and sold to the highest bidder, but the authorities found it out, went in and took such stores. And one firm, Goldberg, Bowen & Company, are just about killed in this town.[43]

Goldberg, Bowen rebuilt in 1909, and that beautiful building still stands today, at 250 Sutter Street.

The Emporium
When Adolph Feist opened the Emporium on Market Street in May 1896, it was the biggest department store on the West Coast. By the first holiday season, the Emporium included one of the city's largest groceries, along with

a butcher shop, fish market, fruit and vegetable store, and a wine and liquor store, all in the basement next to the Grand Restaurant and Grill Room. From the beginning, the grocery department advertised its high-quality foods, proclaiming itself in favor of Pure Food, in contemporary terms. Shoppers visited the "Pure Food Show," which took over half of the great basement. The store emphasized that it sold "Pure-Food Products" at the "LOWEST possible prices," and then flipped that slogan to reassure its patrons: "LOW as OUR prices are, only Reliable Standard Brands are offered." Among those brands were Pure-Food Maple Syrup, Schepp's Shredded Coconut, Crosse & Blackwell's Lucca Oil, Webb's Cocoa, and Bar Harbor Catsup.[44]

Familiar brands were not always the focus. Advertisements did not list brand names for many products, such as canned Baltimore oysters, Norway mackerel, imported herring, mince meat, imported Swiss cheese, pickled butter, eastern cream cheese, canned pineapple, fancy prunes, and fancy New York dried apples. Some ads featured Emporium Blend coffee, the house brand. Heading into the holiday season a year later, the Emporium's full-page ads continued to feature a mix of branded products alongside unbranded items. Heinz's Keystone Chili Sauce, Durkee's Celery Salt, and Aunt Jemima's Pancake Flour did not get more prominent billing than the anonymous fresh nutmegs, egg noodles, smoked beef tongues, canned blackberries, dried cherries, guava jelly, haricots verts, petits pois, truffles, imported Smyrna figs, fancy navel oranges, fancy Salinas Burbank potatoes, and cheap cocoa shells. (People boiled cocoa shells for three hours to make a hot drink with a little chocolate flavor.)[45] Advertisements also emphasized that you could shop either in person or over the telephone and expect prompt delivery by special grocery wagon. And during the fall and winter, the Emporium advertised itself as "The Rainy Day Shopping Place"—wooing cold San Franciscans into a warm, fragrant shopping environment.

Feist quickly lost the golden aura he had during the opening days of his grand experiment. One of the issues that tripped up Feist was turning a tolerant eye to "impure" foods at a time when the Emporium was making a name for itself as a purveyor of Pure Foods. In the first months of 1897, the Emporium was advertising its high-quality "water white" honey, but apparently someone was cutting corners. In March 1897 San Francisco's chief food inspector, James Dockery, launched a campaign against the sale of adulterated foods and streamlined the process of analyzing suspect foods. The city chemist no longer had to determine all the components of a particular sample of food; he could simply demonstrate that he had found some material the consumer would not expect. Dockery embarrassed Feist by proving that despite the Emporium's grand claims to sell only "Pure Food," some of the store's honey had been adulterated. (A San Francisco doctor said that year that every grocery in the city sold glucose syrup flavored with honey in place of the genuine article.)[46] By 1898 Feist was looking for a new line of work. The store's honey was no longer featured in its advertisements. And the Emporium was off to a new start under Frederick W. Dohrmann, who merged the company with the well-established Golden Rule Bazaar and ran the Emporium until 1914.

The Emporium was completely destroyed in the 1906 disaster. A month later, however, the managers had opened a temporary location on Van Ness and telegraphed an immense buying order to the East Coast, confident that the city would be shopping again before long. By July it was assuring *Chronicle* readers that "we practically bring the markets of the world to you"—advertising the riches of those global markets with imported teas and that up-to-date dessert treat: Jell-O.[47]

When the rebuilt Emporium building on Market Street opened in October 1908, prominently located on the first floor next to the expanded grocery department was a new soda fountain and an attractive new delicatessen. The store made a point of having enticing food displays, especially leading up to the holiday season. In early November, for instance, shoppers entered the store and encountered ten thousand pounds of California glaced fruit in fancy wooden boxes—the perfect present for friends and family back east. Later in the month, along with displays of mince meat, black and white California figs, almonds, walnuts, pecans, chestnuts, filberts, and Jordan almonds, the store displayed a new treat known as the Dutch lunch box: "appetizing, unique confections in the form of meat, sausages, sauerkraut, potatoes. . . . These are so entirely new and original that they cannot fail to be appreciated by any one."[48] The Emporium took as its mission not only providing "pure food" to its customers but also finding ever-new ways to make the shopping experience innovative and fun. On his visit in 1915, Pennsylvania native Rudolph Walther was impressed by the glamorous refrigerated cases showing off beef, salmon, and halibut steaks, and he noted the diverse offerings of the delicatessen department:

> Delicacies from all parts of the globe, from the English Chilton cheese down to the Bismarck herring. Salads of all common varieties in bulk, as coleslaw, potato, herring, shrimp, crab, and lobster, with a rich, creamy mayonnaise dressing, all ready to eat. . . . It was a feast for the eyes.[49]

Pure Food Movement

When the Emporium associated itself with Pure Foods in 1896, the movement was growing quickly. The idea was influential on the East Coast in the 1880s, and it resonated in California as early as 1886. The state's agricultural interests were proud of their produce and spoke out against fraudulent labeling:

> The products of the soil leave the hands of their producers in a pure and wholesome state. Before they reach the consumer they are too often transformed by sophistication. . . . The pure article meets its deleterious counterfeit underselling it in the market, to the ruin of the farmer and the destruction of public health. . . . Here we have the greatest variety of useful products, and in many of them we enjoy an absolute monopoly of production. Our wines, raisins, and many of our fruits will never have rivals upon this Continent, and therefore California speaks for the whole country when she demands that her . . . fields and dairies [be] freed from criminal competition.[50]

The Pure Food movement gained strength in the 1890s, spearheaded by the wine, fruit, and dairy industries as well as by private individuals frustrated by high levels of adulteration in the products their families consumed every day.

In February 1895 San Francisco hosted a Pure Food Fair at the Mechanics' Pavilion. The fair was sponsored by the city's grocers and featured samples of milk, cheese, candy, coffee, chocolate, wines, fruits, jellies, and preserves, as well as "other appetizing articles in cans and bottles." Upscale grocers Bibo, Newman & Ikenberg promoted their syrups, brandies, jellies, preserves, and spices as "absolutely pure, home made . . . from the very best and purest selected materials." There were so many offerings that attendees reported cheerfully one could eat "five square meals of pure food" in a single afternoon at the fair, all free with the single admission fee of twenty-five cents. No one noted why one would want to eat five square meals in a single afternoon—but at least it was possible.[51]

Just over a month later, the California Legislature passed the Pure Food Law of March 26, 1895, which provided financial penalties for any adulteration of the state's food and drugs. Later that year, Chief Food Inspector Dockery gained fame using the law to go after vendors selling fraudulent milk:

> Adults . . . adulterate [milk] of their own volition with tea, coffee, whisky, or something else. . . . [But] if the people of San Francisco could realize the dangers to which their babies and children have been exposed from the use of impure and adulterated milk they would vote a statue to the officials . . . [who exposed] the nefarious traffic, and who have determined to stop wholesale child murder.[52]

A few years earlier, the city had cleaned out "Cow Hollow" (north of Union Street, between Polk and Divisadero), ridding the district of its urban dairies and their associated stench after cows were found suffering from tuberculosis and living in unsanitary conditions. Dockery knew San Francisco's residents were primed to believe the worst of their milk.

Now, in 1895, the *Chronicle* ran a series of articles promoting Dockery as a sort of Robin Hood and his assistants as modern highwaymen, stopping milk deliveries to demand: "Your milk or your life."[53] Some of the articles read like adventure tales, with titles such as "He Will Destroy All Impure Milk: Inspector Dockery on the Warpath"; "Dockery Had to Use His Pistol"; "A Merry War on the City's Milk: How the Raid Is Carried On"; and "Tried to Fool Dockery: How He Caught a Tricky Milk Dealer." Dockery cleaned up the milk business, preventing dairymen from selling skim milk or watered-down milk as if it were whole.

Pure bread was the next major push, as journalists looked into the conditions in San Francisco's bakeries, where "perspiring men kneaded bread amid dirt, heat, filth and cockroaches." According to reports, some of the basement bakeries were in "intimate relations with poor sewers." One problem was that San Francisco had so many bakeries, the city found it hard to monitor them, leaving that job to the vagaries of open competition:

> There are those, perhaps, who will regard this as a guarantee of a cheap loaf, but it is not. The home kitchen guarantees low-priced bread, and what the city gets as a result of bakery competition is, in many cases, a deleterious compound. . . . Bread, like water, meat, milk and air, must be pure or the public health suffers. The milk crusade was one of the best reform measures which San Francisco has seen in force for many a day. The bread crusade is not second to it in importance.[54]

Testing bread for purity was harder than testing milk, and small businesses had a hard time proving their bread was clean. In the long run, the Pure Food crusade helped push many of San Francisco's small bakeries out of business and led large bakeries to systematize and mechanize their bread making.

Other products were even harder to control. When the Manufacturers' and Producers' Association convened the "Pure Food Congress" in San Francisco in the spring of 1897, members complained that the law was not applied enough and did not even prevent eastern growers from shipping adulterated fruit for sale on the West Coast. From then on, California's producers devoted more energy to the movement for a national law, which finally came into being as the Pure Food and Drug Act of 1906.

Even as they pushed for state and national Pure Food laws, however, California's farmers worried that outside experts might misunderstand the

technical details of particular agricultural processes. In particular, fruit growers commonly used sulfur to preserve the color and taste of plums, peaches,
and apricots—insisting their dried fruits were nevertheless still "pure." Dairy
interests were eager to use the Pure Food movement to obstruct oleomargarine manufacturers calling their product "butter," even though some milk
producers added bicarbonate of soda to prevent milk from souring, and butter
producers added salt for taste. Agricultural interests tried to maintain their
reputation as allies of the movement while working to exempt common
practices from public scrutiny and opprobrium. Similarly, California's meat
dealers tried to keep out German sausages as contaminated with borax, while
hoping to keep inspectors from looking at how Californian sausages were
made.

The Pure Food movement had a strong presence at the Panama-Pacific
International Exposition in 1915. An organization of retail grocers arranged
for a "Pure Food" ball at the exposition and had the grand idea of inviting
their suppliers to arrange for costumes representing common grocery offerings. The twelve thousand men and women who gathered for the ball at the
Civic Center Auditorium were entertained by "an animated pineapple with a
dizzy asparagus tip doing the fox trot, followed by a graceful bottle of ketchup
and a can of pears doing the hesitation [a fast waltz]."[55] Attractive women
filled out most of the two hundred "living impersonations" leading the Grand
Pure Food March, the capstone of the evening's excitement.

The Cost of Living

During the Gold Rush, the costs of living in San Francisco were variable and
often expensive, as retailers and restaurateurs tried to "mine" the miners. By
the 1880s, however, prices were in proportion with other cities around the
country. And by the turn of the twentieth century, employers complained
that San Francisco was a utopia for workers:

> No other city in the United States is so completely dominated by the labor
> unions today as is San Francisco. . . . Wages [there] are higher than in any
> other city in the world. . . . The cost of living in San Francisco is remarkably
> low. . . . The fuel bill there is small, while vegetable and fruit products are
> plentiful and cheap.[56]

In the twenty-first century, it is hard to imagine San Francisco as a cheap
place to live. But around 1900 the mild climate meant relatively low fuel bills
for city residents, fresh meat and eggs were cheaper than on the East Coast,

and during the winter the Bay Area offered a greater variety of fresh produce at lower prices than East Coast cities.

One complication in studying retail prices of different products is that over the years, people turn to different products, both to keep down costs and to suit their changing tastes. Butter became more expensive at the beginning of the twentieth century, and then less expensive again in the Great Depression. But to understand how that affected the cost of living, one would have to consider how consumers felt about saving money by switching to the new product "oleomargarine." Some people saved money by purchasing "cold storage" butter and eggs—both were sometimes sold at lower prices than the fresh versions and sometimes passed off as fresh by unscrupulous grocers. San Franciscans' grocery bills also changed as people began adopting new canned goods and processed foods, from Post Toasties corn flakes to Van Camp's pork and beans.

Labor Issues

The history of labor actions in San Francisco showed the benefits of standing firm for workers' rights. In 1891 the Manufacturers' and Employers' Association had destroyed the waterfront unions' influence for a decade, although the White Cooks and Waiters' Union managed to maintain its own power over restaurant owners. In 1901 the City Front Federation tried again, shutting down the movement of goods on the waterfront from July 30 to October 2, during the region's prime growing season. The Teamsters and the rest of the City Front Federation were joined by a range of other locals, from beer bottlers and bakery workers to the cooks and waiters' union. Together, they achieved many of their goals and effectively shut down the power of the Employers' Association. Over time, however, employers fought back, working to make food shops, groceries, and restaurants "open shops" (open to union and nonunion workers) rather than "closed shops" (restricted only to union members). By 1906 many restaurants were open shops.[57]

Some grocery stores were so small they had no employees, and yet unions still targeted them for ignoring union-approved working limitations. In 1908 members of the Grocery Clerks' Union targeted small stores that stayed open after 7 p.m. The owners complained that their customers worked during the day and needed to shop in the evenings, but the union pickets appeared in front of the stores the next day with signs saying, "Boycott! This store is antagonizing the Grocery Clerks' Union in its endeavors to maintain fair conditions."[58]

During the Great Depression of the 1930s, about a quarter of San Francisco's workers were unemployed and receiving government relief aid. Activists pushed for greater control of labor issues, from hiring to the length of the workday. A shorter workday would spread jobs around, increasing the number of people who could bring home a paycheck to feed their family. Waitresses' Local 48 picketed almost three hundred restaurants in 1933, demanding a forty-hour workweek in place of the fifty-four hours called for by the National Restaurant Association.[59]

In May 1934 longshoremen up and down the Pacific Coast went on strike, asking for reduced hours, higher pay per hour, and a union-run hiring hall. The shipowners hired scabs and police protection to circumvent the strike and, on July 5, strikers and police fought a bloody battle that left two workers dead. In the aftermath of that brutal conflict, unions called for a general strike. Families stocked up on groceries in preparation for the citywide shutdown, which began on July 16, 1934. Prominent Democrat organizer Julia Gorman Porter remembered the lead-up to the strike:

> Everybody bought a ham and a side of bacon to prepare for the siege. . . . The thing I did was go to Goldberg Bowen and order ham. . . . Of course, the strike was over rather shortly, and for the next six months every time you were invited to dinner, you had roast ham.[60]

Figure 4.4. Men lining up for bread at Compton's store during the 1934 dock strike. Courtesy of the San Francisco History Center, San Francisco Public Library.

During the strike, Teamsters' pickets prevented food trucks from reaching San Francisco. Almost all restaurants closed; union officials did allow Compton's Cafeteria, Maison Paul, and a few other restaurants to stay open and feed the strikers. Grocery stores could not resupply their shelves, although the Labor Council let trucks carrying ice, milk, and bread circulate as usual. Strike organizers were concerned about losing the public's goodwill. By the second day of the strike, fifty restaurants were allowed to open; by the third day, all restaurants were open. California Highway Patrolmen began accompanying trucks of fresh produce to the city, and the National Guardsmen protected the Colombo Market, but by that time they were unnecessary. After three days, the general strike was over.[61] In the aftermath, San Francisco's unions opened up further to people of color, and Waitresses' Local 48 became the largest union of waitresses in the country. By the end of the decade, culinary unions had won an eight-hour day and equal pay for men and women from the city's hotels, lunch counters, cafeterias, ice cream shops, and department store restaurants.[62]

Supermarkets

Supermarkets were born during the Great Depression, growing out of a longer trend toward chain stores. Consumers wanted to buy in larger quantities to save money, and both stores and homes now had refrigerators to store the extra food. Large local establishments such as the Emporium or the Crystal Palace Public Market at Eighth and Market now seemed behind the times. National supermarket advertising campaigns pulled people away from neighborhood shops and toward this new way of shopping. Among the first supermarket chains in San Francisco were Safeway, Mutual Stores/MacMarr, and Piggly Wiggly Pacific, followed by Lucky, Cala Foods, and QFI.[63]

Small grocery retailers in the city could not afford the real estate costs of a San Francisco supermarket. After spending decades building up a reputation for supplying particular ethnic foods for their small neighborhoods, these store owners faced competition from major chains also offering tortillas and jalapeños, matzo and horseradish, tofu and soy sauce. Chinese Americans operated some supermarkets outside San Francisco. In the late 1940s Bill Wong, the eldest son of a Sacramento-area Chinese farming family, opened a supermarket in Berkeley. His United Food Center took up almost a whole block in an African American neighborhood. And in the 1950s Sam Wah You bought eight Hob Nob supermarkets in the South Bay.[64]

The shift to an ever-wider range of products accelerated throughout the twentieth century. Supermarkets in San Francisco were at the forefront of

Figure 4.5.   Safeway grocery store, 1936. Courtesy of the San Francisco History Center, San Francisco Public Library.

this trend. Foreign visitors admired the diversity of products at Safeway's flagship store in the Marina district. And in the 1970s San Francisco–based QFI worked to attract new consumers for these new tastes. Instead of hiding ethnic products in an obscure aisle, for instance, the store promoted tortillas, taco shells, enchilada flavoring, and Spanish rice alongside boxes of Italian pasta in visible end-of-aisle displays. Whether mangoes or bok choy or limes, whether rice vinegar or Red Devil hot sauce, new ingredients flew off the shelves.

Over time, however, mass-market outreach ceded some ground to market segmentation, as supermarkets began specializing in one slice or another of the population. National chains such as Costco, Sam's Club, Whole Foods, and Trader Joe's have gone after the cost-conscious, the health-conscious, gourmets, and environmentalists. Local health food stores such as Rainbow Grocery have had to lower prices to compete. Supermarkets have also sometimes entered low-rent districts, killed the competition, and moved on to chase profits elsewhere, leaving the neighborhood without a convenient grocery store. As one public advocate noted, "*Se fue* is Spanish for 'It's gone.'"[65] In the 1990s, 99 Ranch Market opened supermarkets around the Bay Area (while avoiding high rents in San Francisco proper). The 99

Ranch chain catered to Taiwanese and other Asian Americans, while also playing to new consumer interest in Asian ingredients and techniques. Over the same period, Mi Pueblo Food Centers followed a similar model, catering to Latino shoppers as well as to a broader base of consumers influenced by Latino culture and cuisine.[66]

Farmers' Markets

During the nineteenth century, farmers' markets began to fade away in the United States, as wholesalers and retailers played a larger role between the farmer and the consumer. The Colombo produce market, for instance, was geared toward wholesale buyers from the beginning. Gradually, the covered alleys of most of San Francisco's retail markets gave way to chain groceries and supermarkets.

Public misconceptions and a misleading newspaper campaign drove the first major controversy over farmers selling directly to consumers. In the depression of the 1890s, producers grew frustrated with low payments they received from their wholesaler partners, known as "commission men." Producers believed commission men lied about how much fruit spoiled on the way to market and how much sold at good prices. In 1896, hoping to rectify the situation, farmers made three demands:

1. A new urban produce market closer to rail and steamer to minimize both transport costs and damage to the tender fruits and vegetables.
2. Official oversight of commission sales by the State Board of Harbor Commissioners.
3. Transparency of sales of all perishables sold at the new market.

Farmers did not expect to come to the city and sell their own products. The chairman of the Producers' Committee looked back a decade later and explained that the farmers had always wanted to sell through commission merchants, understanding that "under proper regulation that is the cheapest and best way to sell perishable products." The public, however, assumed the opposite from the very beginning:

> Unfortunately, the theory that producers and consumers would meet in the "free market," where the honest farmer would personally meet the guileless consumer and the two dicker together for truck, was the one picturesque feature which caught the public imagination and served the purposes of the public press. . . . [This] was the only idea that the general public ever got into its head.

Newspapers in both the city and countryside saw the appeal of this notion:

> The city press and people were possessed of the notion that the farmer was yearning for the chance to sell his products at the lowest rate which he could induce consumers to pay. . . . While the country press, on the contrary, seemed equally firm in the conviction that all city consumers desired was the change to pay the horny-handed farmer the very highest prices which he could be induced to accept.[67]

Journalists mangled the farmers' eminently reasonable plan until it became unrecognizable. Instead of a new, supervised commission market on three blocks of very valuable harbor property, San Franciscans believed they were going to get a real farmers' market. Confusion over the farmers' intentions obstructed their efforts to work the state's political machinery in Sacramento. And in the end, the Southern Pacific Company and the Pacific Coast Steamship Company got those blocks "because they attended to their business, and the farmers lost them because they didn't." The public fell in love with the idea of buying cheap produce directly from charming farmers and refused to acknowledge the farmers' real proposal, which kept the farmer on the farm.[68]

Commission merchants maintained their control until the Second World War. In the 1940s Sicilian immigrant and Sonoma farmer John Brucato saw a problem. He began speaking out about the high prices San Francisco consumers paid for produce compared with the low revenue to the farmer. The calculations had changed in fifty years; farmers now could do better by coming to the city themselves. And the war provided an excuse for circumventing the wholesale commission men, who had been letting produce rot in warehouses and colluding to raise prices. Brucato worked with other farmers and the wartime Victory Garden Council to establish a market at Duboce Avenue and Market Street. Almost fifty thousand people showed up that first Saturday, August 14, 1943, eager for high-quality produce. A half-century later, Brucato admitted he had not expected such an overwhelming response:

> There was no plan. Things just happened because it was wartime. . . . People accepted the markets immediately because of the fact that they were getting something fresh. Corn was picked yesterday. If you buy it in the market, it is three or four days old and it has lost most of its vitamins. And the fact that there was a friendliness, too. A lot of farmers developed regular customers.[69]

City residents liked seeing the farmers face-to-face and feeling they were buying from a real person. The battle with wholesalers and retailers continued,

Figure 4.6. Farmers' market at Market and Duboce Streets. Courtesy of the San Francisco History Center, San Francisco Public Library.

but in 1946 the city passed Proposition 17, a ballot measure subsidizing the new farmers' market. Two years later, the market moved to the Alemany Boulevard location where crowds still show up every Saturday of the year, drawn by the market's low prices and lively character.

In 1976 the U.S. Congress passed the Farmer-to-Consumer Direct Marketing Act, which smoothed the way for other farmers' markets around the country. In the Bay Area, the Heart of the City Farmers' Market opened near the Civic Center in 1981, and the Ferry Plaza Farmers' Market opened in 1992 (run by CUESA, the Center for Urban Education about Sustainable Agriculture). A dozen other farmers' markets have opened in San Francisco and about two hundred more around the Bay Area, including one in West Oakland that aims to connect black farmers to the black community. As during Brucato's day, vendors at farmers' markets are supposed to demonstrate to the County Agricultural Commissioner that they are farmers and only selling their own fresh produce. The Ferry Plaza market irritates some city residents with its higher prices, prepared foods, and crowds of tourists, but many San Franciscans appreciate the gourmet local and organic offerings at that site, as well as CUESA's emphasis on education and sustainability. With the arrival in the region of many Southeast Asian farmers, consumers began to see farmers' markets as an opportunity to discover new fruits and greens. Community Supported Agriculture (CSA) programs provided a similar service, with farmers' delivering familiar and unfamiliar produce to many Bay Area residents in weekly boxes.[70] Farmers' markets, CSA programs, and grocery stores with organic offerings all helped encourage home cooks who were customizing the new California Cuisine movement to their own tastes.

CHAPTER FIVE

~

Famous Restaurants

From the very first days of the Gold Rush, San Francisco earned a reputation as a restaurant town. Ships brought exotic ingredients from all over the world, as well as people used to many different cuisines. Early reports emphasized the diversity of restaurants, the fact that almost all meals were either eaten in public or brought home ready-to-eat, and the vast amounts that San Franciscans were spending on food.

At the same time, many doubted that restaurant customers were getting what they paid for. Contemporaries raved about having so many new cuisines to explore, but also revealed anxiety about having to assess the value and wholesomeness of unfamiliar dishes. French-run and Chinese-run restaurants predominated in the first year or two of the Gold Rush and elicited similar concerns about what was hiding under their sauces. Hubert H. Bancroft disparaged Chinese dishes as "deceptive mixtures, not unlike that of the French," and criticized both cuisines for obliterating "the original taste or essence of a food" with their condiments, "processes," and sauces.[1]

As sophisticated cosmopolitans, San Franciscans wanted to spend their money on great meals. But as no-nonsense Americans, they also did not want to be duped into eating something dubious. That fear of overpaying for unhealthy or unpalatable meals meant diners appreciated reliable restaurants. Restaurateurs worked hard to build and maintain relationships with their clientele over decades, even if that meant straining to link a new restaurant back to an earlier restaurant with a solid reputation. A chef in common, a beloved waiter, a shared location, even just a shared ethnic background with the earlier owner—anything could be used to woo paying customers. From the Gold Rush to the later era of California Cuisine, San Francisco cherished restaurateurs who stressed continuity and community as well as new tastes and trends.

Tadich Grill

John Tadich of the Tadich Grill was a great raconteur and entertained customers with anecdotes of the Gold Rush days in San Francisco. On slow afternoons, Tadich traced the origins of the restaurant to a ramshackle tent on Long Wharf in 1849. He admitted openly that he was relying on stories he heard from others and regretted that so few records were kept of the Croatian community's early years in San Francisco.[2]

According to these stories, in 1849 Croatians Nikola Budrovich, Frano Kosta, and Antonio Gasparich began serving coffee and some food from a tent on Long Wharf. By 1851 the Wharf was renamed Commercial Street; city directories that year listed Nicholas Bennett, Frank William, and Anto Gaspar running a coffee stand at 27 Commercial Street. Business was apparently better with anglicized names. As the city redeveloped the waterfront, the coffee stand moved to the New World Market. There it became known as the New World Coffee Stand under a succession of (mostly) Croatian owners.

In 1871 sixteen-year-old John Tadich arrived in San Francisco and was quickly integrated into the tight-knit Croatian community. He worked at his uncle Nikola Buja's coffeehouse and saloon at 605 Davis Street and then at another Croatian saloon at Pacific and Sansome. In 1876 he began to work as a bartender at the New World Coffee Stand, then co-owned by Croatians Sam Bećir and Frederick Bralich and French immigrant Eugene Masounette. Tadich's new workplace would change his life and change San Francisco's culinary history.[3]

In 1882 Masounette and Bećir renamed their business The Cold Day Restaurant after an incident involving tax assessor Alexander Badlam, a nephew of the colorful pioneer businessman Sam Brannan. Badlam had assured voters of his reelection with the phrase "It's a cold day when I get left"—meaning it was unlikely he would be left out in the cold. Defeated, he sought comfort at the New World Coffee Saloon, then at 221 Leidesdorff. Masounette and Bećir teased Badlam by incorporating his campaign slogan into the restaurant's new name. The "Cold Day Restaurant" resonated with San Franciscans in search of warm comfort, and the venue soared in popularity.

By 1888 Tadich had bought out Masounette and Bećir. Tadich kept the name and location while Masounette and Bećir went on to open another restaurant. After the 1906 earthquake, Tadich joined forces with another Croatian restaurateur, John Sutich, to re-open the Cold Day Restaurant on Pine Street. In 1912 their partnership ended. This time, Sutich kept the name and the location, while Tadich moved to 545 Clay and opened a restaurant called the Tadich Grill: The Original Cold Day Restaurant. A reasonable person might think that was the start of today's Tadich Grill. But that reasonable person has no understanding of marketing.

Tadich delighted in recalling the early days of his tenure at the New World Coffee Stand, when the neighborhood bubbled over with publishing houses, and notable writers would stop in to have dinner and chat. San Franciscans enjoyed Tadich's stories, just as they enjoyed his crab salad, fried sand dabs, and sweetbread patties in Madeira sauce. Many customers followed him from 221 Leidesdorff to Pine Street and then to the Tadich Grill on Clay Street. They supported his vision of continuity stretching from a tumbledown tent on Long Wharf to the Tadich Grill.

By 1934, three Croatian brothers took over—Mitch, Tom, and Louie Buich, all former employees—and the Tadich Grill has been in the Buich family ever since. Louie Buich began cooking seafood over the restaurant's mesquite broilers, grilling fish with techniques long established in Croatia. (Croatians have cooked a lot of San Francisco's meals, due in part to their native skill with fish.)[4] Much as the Tadich Grill always bought its bread from the Parisian Bakery, the restaurant has had the same mesquite supplier for more than one hundred years: Lazzari Fuel, an old San Francisco company. Mesquite charcoal cooked the fish quickly, adding flavor without overwhelming the fish. Customers loved the taste, and the Tadich Grill solidified its reputation for delicious seafood, from abalone to California yellowtail, by way of finnan haddie, octopus, petrale sole, prawns, salmon, sea bass, swordfish, trout, turbot, and many more.[5]

By any measure, Tadich Grill has been a San Francisco institution for more than one hundred years. Longtime local columnist Herb Caen loved to go there and brought anyone who mattered in San Francisco politics. Willie Brown, longtime California state assemblyman and later mayor of San Francisco, first went to Tadich with Caen and did much of his political arm-twisting in its booths. San Francisco mayor Joseph Alioto grew up in the neighborhood, working at his father's wholesale fish market, the San Francisco International Fish Company. "Tadich was in the center of the three leading wholesale fish markets," remembered Alioto: "The fish that was filleted at five o'clock in the morning would be served at Tadich for lunch."[6]

The Tadich was unpretentious and fond of traditions. Prices stayed low, everyone but Tony Bennett had to wait for a table, and sawdust was on the floor until 1967 when Steve Buich moved the restaurant from Clay Street to its current location at 240 California Street.[7]

Poodle Dog

In 1858 a recent French immigrant took out an ad in the *Daily Alta California*:

> N. F. Richit Announces to the Public that he has opened the Union Rotisserie and Restaurant on the southwestern corner of Washington and Dupont Streets, under the direction of Mr. Marchand, who is well known in San Francisco for his skill in the Culinary Art.[8]

Nicholas Richit had come from New York during the Gold Rush years. Prior to opening the Union Rotisserie, he and his wife, Eugenie, owned and ran the Union Grocery on that same southwestern corner of Washington and Dupont (now Grant). The best evidence suggests that Richit's 1858 Union Rotisserie and Restaurant was the first incarnation of a famous San Francisco restaurant, the Poodle Dog.[9]

Nicholas and Eugenie Richit worked hard to make their new restaurant stand out from the many other contemporary French restaurants. Together with their French chef, Edward Marchand, they put in long hours at the restaurant. They briefly brought in another partner, Augustus Esnault, but he left after a few months. In late November 1859 a terrible fire destroyed the building, and Richit found it hard to bounce back. In 1860 Marchand was running the Union Rotisserie with help from his wife, who liked to bring her French poodle to the restaurant—San Franciscans embraced the dog and began referring to the restaurant as the "Poodle Dog."[10] In February 1861 the restaurant burned down again, and the next year Richit died. Marchand went on to other ventures, first to the Union Hotel on Kearny, and then opened his own elegant French restaurant, Marchand's. But those few years with the poodle had made an impression on San Francisco.

Other French rotisseries began to pop up around the neighborhood. In 1863 Richit's former partner Augustus Esnault teamed up with Peter Bajo and Alexander Finance to re-create the Poodle Dog. They started small, with just a coffee saloon at 837 Dupont, but they brought staff from the Union Rotisserie; by 1865 they had a French Rotisserie up and running at 825 Dupont, just down the block from the original location. At the time, it was officially called "Etienne & Co." But customers called it the Poodle Dog because so much of the flavor of the beloved old restaurant remained.[11]

In 1869 the *Chronicle* made racy jokes about the Poodle Dog and the indecencies taking place in private rooms upstairs:

> Private Rooms and Suppers at the "Poodle Dog": Ladies, have you ever heard of the "Poodle Dog?" That is the charming and seductive haunt where your husbands spend their evenings when they deny themselves the pleasure of your society on the pretense that they have to attend "the Lodge" or a "mining meeting." At the "Poodle Dog" they indulge in elegant and expensive suppers, with the choicest cooking and the rarest wines. . . . San Francisco is full of fast women, who lay their snares for pecunious married men.[12]

By 1873 Alexander Finance had moved to Paris. An auction house informed San Franciscans, "on the premises, No. 825 Dupont street, between Clay and Washington, we will sell the Stock and Fixtures of the Poodle Dog Restaurant."[13] Two ambitious Poodle Dog employees seized the opportunity. François Péguilhan and Jacob Stork reopened the restaurant at 445 Bush Street, on the corner of Dupont. They even officially listed it in the city directory as the Poodle Dog Restaurant.[14] In 1875 Bret Harte's *Overland Monthly* magazine published a restaurant scene set at the Poodle Dog:

> The table glittered with . . . crystal, gold, silver, and Japanese china; with red Burgundy and Bordeaux, yellow Clicquot, and Frontignan [a sweet white wine]; with truffles and terrapin, spiced salads and meats, flaky pastries, tempting fruit, and fanciful confectionery. A giant bowl hollowed in ice held a claret punch.[15]

The restaurant was gaining fame even beyond San Francisco. A French tourist visiting San Francisco in 1879 recommended the Poodle Dog as "the best French restaurant in the city."[16] Péguilhan had worked at the Poodle Dog since at least 1871 and may have been part of the group that opened the 1865 restaurant on Dupont. He and Stork continued to build the restaurant's reputation for fine cuisine and savvy marketing until 1880 when Leopold Ligon took over Péguilhan's share.

In 1895 Antonio Blanco bought the Poodle Dog with an eye to modernizing it. He moved the Poodle Dog to the northeast corner of Eddy and Mason Streets, in a brand-new six-story building built with every new restaurant amenity. It was designed to be fireproof in a city often struck by fire. The basement had not just a vast wine cellar but also vegetable rooms, bottling rooms, state-of-the-art refrigerators, and a laundry. The kitchen on the first floor had broilers, heaters, warmers, driers, iceboxes, and "patent dishwash-ers." The range was enormous and double-sided, with one oven on the front and another accessible from the rear: "the only one of its kind in this country . . . a thing worth seeing."[17]

The main dining room of the new building had Venetian chandeliers, French wallpaper, and rococo furnishings—and women were not allowed in after 5 p.m. without a male escort. The top floor hosted grand events, with two banquet halls and a separate sixth-floor kitchen. In between was a bit more risqué. The second floor had private dining rooms seating six to fifteen people, each room with its own piano. The next three floors featured glamor-ous suites with polished brass beds and marble bathrooms. The brochure put out by the Poodle Dog at the turn of the twentieth century did not elaborate on the purpose of those suites—the people who needed to know already knew.

The restaurant attracted wealthy diners,including businessmen who wanted a discreet but luxurious place to take their mistresses for leisurely meals:

> Price was no object to the Poodle Dog's patrons. They paid 60 cents for a New York cut steak, 70 cents for a T-bone steak—without blinking. Let's shoot the works on a seafood dinner, start off with Russian caviar, go on to clear green turtle soup, terrapin or lobster a la Newberg and enjoy the finest champagne.[18]

At the turn of the twentieth century, the Poodle Dog's menu was seventeen pages, with many preparations of seafood, beef, and seasonal game: duck, pheasant, valley quail, sage hen, deer, and bear.[19]

Not everyone loved the new, modernized Poodle Dog. The others fre-quented the Old Poodle Dog Restaurant, which stayed on Bush Street under different management. As the Chamber of Commerce noted after the fact, "There were patrons for both, and both were excellent."[20] In 1906 both burned down in the great fire. The Old Poodle Dog reopened that fall, at 824 Eddy Street. Blanco first opened a restaurant called "Blanco's," which shone during the reconstruction period and was considered "by far the best restaurant in the city."[21] A few years later, he reopened the Poodle Dog restaurant at 117 Mason Street. By that time, however, the Old Poodle Dog owners (Jean B. Pon, Pierre Carrere, and Calixte LaLanne) had joined forces with two other respected old restaurants, the Bergez Restaurant and Frank's Rotisserie. The merger resulted in a beloved restaurant known as Bergez-Frank's Old Poodle Dog, located at 415 Bush Street, at the entrance to tiny Claude Lane.

Old Poodle Dog Restaurant

824 EDDY STREET   SAN FRANCISCO

(Formerly Bush Street and Grant Avenue)

LUNCH

Cold Meats and Salads

DINNER

With Wine $1.00

SOUP

Clam Chowder   Sorrel

Consommé

FISH

Skate Fish Brown Butter   Salmon

Filet of Sole Tartar Sauce   Sand dab

Silver Smelt   English Sole

French Imported Snails .50   Pompano .50

ENTREES

Braised Chicken à la Casserole

Veal Tongue Spanish

Shirred Eggs à la Meyerbeer

Kidney Sauté   Blood Pudding and Sausage   Liver Brochette

Calves' Head   Lamb Chops   Calves' Brains

Boiled Pig's Feet   French Andouillette

Rump Steak

Canvasback 1.75   Mallard 1.50   Sprig 1.25   Teal .65

Assorted Desserts

30 November 1906

Figure 5.1. Old Poodle Dog menu, 1906.

Prohibition brought an end to these incarnations of the Poodle Dog, but in 1933 LaLanne opened the Ritz French Restaurant at 65 Post Street. When Calixte LaLanne died in 1942, his son Louis renamed the restaurant the Ritz Old Poodle Dog. Louis hired a great French chef, Henri Tarnac, who impressed Doris Muscatine with his calf's head vinaigrette, sweetbreads, braised oxtails, and tripe à la mode de Caen (sealed and oven-steamed for two days, no relation to Herb Caen). Muscatine was, however, a bit perplexed by Tarnac's insistence that garlic was only suitable for snails and Caesar salad. Today's San Franciscans remember that restaurant on Post as the authentic Poodle Dog restaurant.[22]

Campi's

In 1859 Giacomo (James) Campi and John Mauletti opened a coffee stand and restaurant on the corner of Sansome and Merchant. Swiss and Italian, respectively, they welcomed customers in many languages and were known for being able to serve a familiar taste of home to San Franciscans of almost any national origin. As Campi's gained a steady clientele, the restaurant began promoting French and Italian cuisine. In 1867 Campi's moved to the 500 block of Clay Street, where it stayed until the 1906 earthquake. In its new home, Campi's started advertising specials in the Chronicle: "Maccaroni in Neapolitan style, Risotto in Milanese Style, at all hours."[23] A few years later, non-Italians were also delighting in the Italian dishes served at Campi's. In his guidebook to the city, Benjamin Lloyd wrote:

> For the last few years Campi's Italian restaurant has enjoyed high repute among those of epicurean tastes, on account of its excellent cuisine. . . . The proprietors are Italians, and also all the waiters and other employees. It is the Italian cookery that attracts its customers. . . . From 1 to 2 o'clock p.m. . . . it is almost impossible to obtain a seat at Campi's.[24]

The restaurant also educated San Franciscans about the importance of olive oil in Italian cuisine.

Despite its full tables at lunch, Campi's struggled to turn a profit. In 1871 Campi and his partner, Natale Giamboni, were looking to sell the restaurant; later that spring Campi died in a tragic fall at Yosemite. Giamboni took the restaurant in hand afterward and brought in new partners, including former waiters Ildefonse Cuenin and Luigi Geminiani.[25] In 1880 Campi's ran a suggestive front-page advertisement to entice customers with expensive tastes: "Wives are angry when husbands dine at CAMPI'S, 531 Clay St., for they cannot thereafter be pleased at home." The management then ran ads appealing to day-trippers, back late after spending hours at the shore or hiking the hills: "For the accommodation of excursionists, we will hereafter keep our establishment open every evening until 10 o'clock." They even ran ads promoting the restaurant to people with indigestion: "If you are dyspeptic, try an Italian dinner at Campi's Italian restaurant."[26] The marketing campaign did not solve the problem. By the end of the year Campi's was still deeply in debt.

Still, the restaurant remained popular, perhaps due in part to its pricing policy:

> All dishes were priced at a bit apiece, except for wild duck, which was two bits. A . . . bit could be either ten or fifteen cents, according to circumstances. For thirty-five cents, for instance, one got three dishes.[27]

For fifty cents, one was served a full dinner with a bottle of Bordeaux. When the American feminist writer Caroline Dall was visiting San Francisco in 1880, she ate at Campi's for lunch one Sunday. She described a plain-looking space—not a mirror or chandelier to be found—just long tables covered with coarse gray cloth, "nearly full of men, women and children, chattering in all conceivable tongues." Worcestershire sauce, mustard, pickles, and sugar sat in the middle of each table. Dall was happy to note that the jars looked clean. The waiter gave them each a goblet filled with ice, so they could enjoy their "good California claret." They started with French bread and a seafood or potato salad, served on lettuce: "Delmonico never seasoned it better." Then she had a watery bouillon (her tablemate's tomato soup was better), followed by mussels, "hot from the fire . . . very sweet." Her favorite dish was "a good-sized plate of vermicelli, well-cooked, and dressed with beef gravy, seasoned with cheese, Worcestershire, and tomato." The chicken stew was tasty; even though the chicken was tough, it added flavor to the rich stew of pattypan squash and tomatoes, accompanied by oyster patties smothered in cream. Next, they tried fried cream, a dish they had never heard of: "A nicely-made 'Italian cream,' shaped in a 'brick,' is cut into pieces . . . dipped in oil and then sprinkled thickly with cracker-crumbs." Drenched in peach brandy, it burned brightly, browning to the diner's taste. That was followed by coffee and dessert: custard pies, green-apple marmalade pies, and pastries with whipped cream.[28]

Owner Giamboni became known as the "King of Hosts," charming his customers and remembering their tastes from one visit to the next. He managed the restaurant until his death in 1884; former waiter Ildefonse Cuenin then took over Campi's. After the great 1906 fire, Campi's reopened, first on Ellis and then Market. The owners continued to serve fine Italian cuisine, albeit to smaller crowds. By 1917 the restaurant had gone out of business. In 1921 the retired opera singer Delbora Campagnoli opened a restaurant on Broadway with the name Campi's. She made a go of it for a few years, earning praise for her Italian cooking, but by 1925 Campi's was gone for good.[29]

Cliff House

The Cliff House was built on a beautiful lookout over the ocean with a view of the hills north of the Golden Gate. In June 1863 it opened to the public—one of San Francisco's first tourist spots. Ever since, San Franciscans and visitors alike have enjoyed driving out to see the seals sunning themselves under the Cliff House, or an ocean storm crashing onto the rocks. At first, people rode there by themselves, sometimes racing their horses along the road. They would order breakfast on the way down to the beach, and come up after a refreshing walk to enjoy quail, spring chicken, or a tenderloin steak.

Mark Twain's first trip to the Cliff House left him enthusiastic: "the appetite is whetted by the drive and the breeze, the ocean's presence wins you

Figure 5.2. Cliff House and seal rocks. Courtesy of the San Francisco History Center, San Francisco Public Library.

into a happy frame, and you can eat one of the best dinners with the hungry relish of an ostrich."[30] He heard that it was even better in the morning, when it was less crowded, so about a week later he determined to visit the Cliff House for breakfast. He and his stockbroker friend would rise at 4 a.m. to enjoy the open road, the sunrise, and, with luck "a vision of white sails glinting in the morning light far out at sea." But luck was not with them:

> The fog was so thick that we could scarcely see fifty yards behind or before . . . a while, as we approached the Cliff House, we could not see the horse at all, and were obliged to steer by his ears . . . could scarcely see the sportive seals out on the rocks, writhing and squirming like exaggerated maggots, and there was nothing soothing in their discordant barking. Harry took a cocktail at the Cliff House, but . . . I yearned for fire, and there was none there. . . . We were human icicles.[31]

Needless to say, he did not recommend the excursion. He concluded: "If you go to the Cliff House at any time after seven in the morning, you cannot fail to enjoy it—but never start out there before daylight." Other bons vivants headed to the coast after an evening of drinking, continuing the celebration at the Cliff House all night with whiskey sours and champagne cocktails. They would stay for an early breakfast featuring white wine instead of coffee and a profusion of excellent dishes, including terrapin, frogs, and snails:

> The meal ends with a California fruit offering—strawberries, raspberries, apricots, enormous peaches, fresh figs and mangoes, all produced near San Francisco. . . . The mango, a little larger than a hen's egg, tastes somewhat like the banana, but has a richer, tropical flavor.[32]

In addition to its beloved breakfasts, the restaurant advertised elegant suppers "in any style": lunches for wedding parties or clubs; and "moonlight parties," held in private rooms.[33]

By 1869 horse-drawn omnibuses to the coast, along the Point Lobos Toll Road, later called Geary Boulevard. In 1881 the German-Jewish real-estate investor Adolph Sutro acquired the property and many acres around, and in 1888 he built a steam train line to bring San Franciscans to the coast. In 1894 he sold to Southern Pacific, which promptly doubled the fare. Sutro fought Southern Pacific to reduce the fare charged from ten cents each way back to five. By standing up to the railroad monopolists, Sutro became popular enough to be elected mayor of San Francisco.[34] That same year, 1894, the Cliff House burned down. Sutro rebuilt it to vaguely resemble a romantic French chateau, but made out of wood. Theodore Roosevelt enjoyed lunch there in May 1903, asking for a second serving of the filet of sole prepared "à la Cliff House."[35] The building survived the 1906 earthquake but burned down again a year later, only to be rebuilt once again by Sutro's daughter—this time in a neoclassic style made of reassuring concrete.

The Cliff House, with its emphasis on drunken escapades and boozy breakfasts, had a hard time coping with Prohibition. It closed at first; when it reopened in December 1920, the restaurant suggested guests might enjoy soft drinks such as Jackson's Napa Soda and ginger ale, products that were "typical and exclusive of San Francisco," or they could explore the Cliff House fountain menu:

> dozens of delicious drinks based on pure fruit and pure fruit syrups and flavors . . . furnished by the Magnus Fruit Products Company, 301 Howard St., the San Francisco firm whose products are becoming nationally known and are the acknowledged leaders in their respective lines.[36]

As part of the restaurant's new focus on local California products, promotional pieces mentioned that the Cliff House would feature only "California fruits," supplied by Charles L. Goetting & Sons of the California Market. Among those California fruits, the restaurant reserved a prominent place on the Cliff House menu for world-famous Del Monte canned fruits and vegetables.

The Cliff House also promised a sweeping range of tasty delights: "hot light muffins," for instance, at its afternoon teas. These muffins were a precursor of the famous Cliff House popovers of later decades—light, almost hollow rolls made from an egg batter, puffed up with steam. Afternoon teas also featured lobster, crab salad, and chicken à la king. Dinners during Prohibition offered "something absolutely new, so characteristic as to give the place a unique distinction among bon vivants: . . . whole boned stuffed squab chickens." Terrapin, trout, and oysters were served fresh from San Francisco's premier suppliers, and Boudin Brothers baked the restaurant's bread: "those delicious French loaves and rolls so indispensable to the epicure." Those who just wanted a snack could have "Exposition frankfurters, the best on the market." The experiment was short-lived. In 1925 the management shut down everything but a coffee shop to wait until they could serve alcohol again.[37]

By the 1950s, the restaurant was known for its drinks again, along with courteous service, a fine mixed grill, and those popovers. As brunch became popular across America, the Cliff House brunch became a San Francisco fixture. Diners enjoyed omelets and seafood specialties from shrimp beignets to crab cakes to cioppino. Over the years, however, many people have said that one goes to the Cliff House for the view, not the food. Herb Caen liked to complain about the Cliff House:

> I'm all in favor of the Cliff House, but I must say that the current edition leaves something to be desired. The Cliff House, to play its proper role in the mystique of San Francisco, should be all red plush and crystal chandeliers.[38]

But, then, complaining that the Cliff House does not live up to one's expectations is a fine San Francisco tradition, going all the way back to Mark Twain.

Palace Hotel

When the Palace Hotel opened in 1875, the media heralded its kitchen as destined for greatness. A banquet in 1876 for Senator William Sharon, one of the hotel's owners, featured elaborate French dishes such as a consommé served with a savory custard; salmon glazed and covered in truffles, quenelles, and crayfish; and boudin blanc à la Richelieu. That last dish started with chicken breasts, pounded and forced through a sieve, then poached, then stuffed with "lightly fried onion, truffles and mushrooms with butter and gravy," simmered and dished up with onions, small quenelles, and truffles.[39] Quenelles and truffles dressed up many a dish in the 1870s.

A dozen years later, the Palace put on a banquet for visiting New York City firefighters, serving "Mock Turtle à l'Anglaise" (calf's head soup made with sherry); broiled quail à la Colbert (with meat glaze and truffles); baked ham in a sweet champagne sauce; and many other dishes, as well as stuffed potatoes, mashed potatoes, and succotash. Palace chefs had begun adding American flavors alongside classic French preparations.[40] In 1891 guests honoring President Benjamin Harrison's San Francisco visit enjoyed Sacramento Salmon à la Regence; filet of beef à la Richelieu; and a salmi of snipe with truffles, among many other elaborate dishes. The reporter noted that the banquet organizers served only California wines, except for the champagnes.[41]

The Palace Hotel was built around a Grand Court, open to the sky and serving, originally, as an elegant way for guests to arrive by carriage. Around 1900, hotel management had a different vision for the court. Vehicles stayed out on the street, and the open court became a lounge known as the Palm Court. In the 1906 disaster, the hotel burned down. After reconstruction,

Figure 5.3. Garden Court, Palace Hotel. Courtesy of the San Francisco History Center, San Francisco Public Library.

the Palm Court began to be used for large banquets, starting with the one celebrating the reopening of the Palace itself in 1909. Seven hundred of the city's most prominent men came to honor the rebuilding of this San Francisco monument. Among those in attendance at the feast were many who had seen the opening of the old Palace in 1875. They enjoyed green turtle soup "a la Ralston" (a nod to William Ralston, the hotel's founder), "Noisettes d'Agneau, Nouveau Palace" (boneless, rolled lamb filets), and Terrapene à la Sharon (named for Senator Sharon).[42] The Palm Court was later renamed the Garden Court and remained a popular venue for banquets.

More ordinary meals took place in different restaurants at the Palace over the years, including a businessman's lunchroom, a ladies' dining room, a café, a supper room lined with tropical palms, and the famous Palace grillroom. By the mid-1890s, the Palace's grillroom claimed a reputation as the most stylish restaurant on the West Coast:

> The Palace grill room is the place, the *fin de siècle* in cafe, restaurant, cui-
> sine—call it what you please—in America. It has no peer, and the zest of the

grill room steak would bring Chateaubriand from his grave—if it he knew of it. The grill room . . . has become the great trysting place of society—of the eating world.[43]

Besides a wealth of Palace restaurants from which to choose, many guests also ate luxuriously in their rooms:

> The table d'hôte and restaurant of the house are not exclusive enough for many of the rich families who live at the Palace, and . . . quite a number of private dinner-rooms were arranged *en suite* when the hotel was built. . . . The orders were given from the bill of fare and so much a day was charged for each member of the family. . . . It was very popular with those who had the opportunity to enjoy it. . . .
>
> One of the families who enjoyed these special privileges was composed of one of the richest men in the city with his wife and daughter. They paid for their apartments and at the rate of $2 each a day for meals. . . . For breakfast two fowls were generally ordered . . . with two porterhouse steaks, chops, fish, eggs, fruit, etc. . . . At dinner nothing but full joints were served, and the table would groan with all the edibles which the ample menu of the Palace afforded. Generally additional covers were laid [for friends]. . . . Manager Livingston opened his eyes at this, and wider than ever when he found that two or three elaborate stands of fruit were sent to this particular family every day. He found he was sending $18 or $20 worth of food to these people daily for which he was only receiving $6. . . . Almost every private dining-room in the house was being worked the same way.[44]

When hotel management put an end to this practice, the privileged families threatened to move out and boycott the Palace, but the hotel, aggrieved, stood firm.

The hotel weathered other crises as well. During Prohibition, the famous Palace Hotel bar became an ice cream shop, which brought a new audience for the bar's famous Maxfield Parrish mural, "The Pied Piper of Hamelin."[45] After the bar began serving alcohol again, ice cream lovers moved to the hotel's new café, Lotta's Fountain, named for the public water fountain at Market and Kearny donated by the famed San Francisco actress Lotta Crabtree. During World War II, the Garden Court featured many grand banquets. The official banquet for the opening of the United Nations in 1945 was held there, and that fall the city's first postwar social season opened with a splash as an audience of fourteen hundred welcomed fifty debutantes in that beautiful open space.

A succession of prominent chefs combined classical training with an appreciation of local offerings. The first chef, Jules Harder, came from

Figure 5.4. Bar of the Palace Hotel. Courtesy of the San Francisco History Center, San Francisco Public Library.

Delmonico's in New York to lead the Palace kitchens. Proclaiming himself a successor to the great French cook Carême, in 1885 Harder proposed a series of books on American practical cookery. Only the first volume was ever published, dealing with American vegetables from artichokes to wood sorrel. In the preface, he instructed cooks to avoid preserves and select the freshest seasonal vegetables for the table.[46] Harder's successor, Fred Mergenthaler, also made it his mission to publicize California's excellent local products:

> California oysters are beginning to be very much liked. California oyster cocktails are now famous all over the world. The canvasback, mallard and teal that are found here are magnificent. The same may be said for the Mongolian pheasant, quail and other birds. As for vegetables, ours is the richest market in the world.[47]

Ernest Arbogast took over in 1902 and quickly made a name for himself with his California oyster omelet, relying on the local oyster's intriguing coppery taste. He went on to create numerous other dishes, including Oysters Kirkpatrick (baked in a spicy tomato sauce and topped with bacon and parmesan

Carte du Jour

GARDEN COURT DINNER

MONDAY, APRIL 28, 1947

HORS D'OEUVRE

Canape of Caviar 2.00   Crab Legs, Avocado Gourmet 1.85   Jumbo Olives 35

Canape Anchovies and Caviar 1.50   Celery Victor 85   Celery 35   Half Avocado 60

OYSTERS

Blue Points on Half Shell 1.40   Fried Toke or Blue Points 1.65   Fancy Pan Roast 2.00

COCKTAILS

Avocado and Grapefruit 60   Sea Food 80   Olympia Oyster 1.20

San Francisco Bay Shrimps 60   Fresh Fruit Cup 50   Crab 90

SOUPS

French Onion Soup with Cheese Croutons 30   Consomme en Tasse, Alphabet 30

Mock Turtle 40   Gumbo 35   Petite Marmite 50

Cold: Creme Vichyssoise 40   Creme Colony 40   Consomme 30   Madrilene en Gelee 40

FISH

Poached Filet of Rex Sole, Chauchart 1.50

Broiled Supreme of Silver Salmon, Bonanza Inn 2.00

Half Broiled Rock Lobster, Parsley Butter, Julienne Potatoes 2.25

California Gulf Prawns on Toast, Newburg 1.75

ENTREES

Shirred Eggs with Lamb Kidney, Meyerbeer 1.50

Breaded Pork Tenderloin, Tomato Sauce, String Beans 2.00

Half Milk-Fed Chicken Saute, Financiere 1.75

Roast Rack of Spring Lamb, Bouquetiere 2.25

Boiled Corned Ox-Tongue, Raisin Sauce, English Spinach 1.50

Breast of Capon on Rice Under Bell, Hongroise 3.25

Baby Veal Cutlet Saute with Fresh Mushrooms 1.75

ROASTS

Roast Prime Rib 2.00   Roast Turkey 1.50   Roast Spring Chicken 1.70

CHARCOAL BROILED

New York Cut 2.75   Tenderloin 3.00   French Lamb Chops 2.00   Veal Chop 1.50

VEGETABLES

Asparagus 50   Asparagus with Hollandaise 75   Artichoke 50   Artichoke with Hollandaise 75

Spinach 35   String Beans 45   Green Peas 45   Baked Potato 35   Au Gratin 40   Long Branch 35   Lyonnaise 35

SALADS

Lettuce 45   Belgium Endive 65   Green Goddess 70   Fresh Fruit 85

Palace Court Chicken 1.55; Shrimp 1.20; Crab 1.65   Chicken 1.60   Shrimp 1.50   Crab 1.95

The above salads are served with French Dressing or Mayonnaise

Thousand Island 25   Rocquefort Dressing 25   Green Goddess Dressing 35   Russian Dressing 50

DESSERTS

Boston Cream Pie 30   Cherry Pie 30   Apple Pie 30   Strawberry Cream Pie 30

Regina Pudding 30   Angel Food Cake 30   Apple Tart 35   Strawberry Short Cake 30

French Pastry 30   Old Fashioned Strawberry Shortcake 65   Small Cakes 30   Baba au Rum 65

ICE CREAM

Coupe Nesselrode 60   Tutti Frutti Parfait 50   Orange Sherbet 40

Bombe Palace 60   Pineapple Sundae 60

Raspberry Water Ice 30   Frozen Egg Nogg 80   Chocolate Roll Glace 60

Strawberry, Chocolate, Vanilla or Mocha Ice Cream 40

CHEESE

American 35   Swiss 40   Monterey Cream 35   Roquefort 50   Philadelphia Cream 35

Brie 35   Liederkranz 40   Yami Yogurt Continental 50   Camembert 40   Cottage 35

COFFEE, TEA, ETC.

Coffee, Pot 30   Kaffee Hag 40

Palace Hotel Special Coffee 45   Buttermilk 15   Horlick's Malted Milk 40

Special Bottled Milk 15   English Breakfast, Ceylon or Orange Pekoe Tea 30

Ry-Krisp, Toasted Rolls or Melba Toast 20

The Palace Hotel

Edmond A. Rieder, General Manager

In addition to the prices quoted above there is a charge of 2½% for Sale Tax.

Figure 5.5. Palace Hotel menu, 1947.

cheese, and named for John C. Kirkpatrick, longtime manager at the Palace).
Less well-known today are Arbogast's deliciously creamy terrapin, cooked in
its own shell; and his corn pudding "à la Sultan," flavored with cinnamon,
mint, bananas, dates, green tea, and rum. In 1923 the Palace's head chef,
Philip Roemer, invented the creamy herbal salad dressing known as "Green
Goddess," in honor of the eponymous hit play of that season. Lucien Hey-
raud made his mark in the 1950s with a return to traditional French dishes:
"all truffles and foie gras," in Herb Caen's words. In their diverse ways, these
chefs all maintained the Palace's strong focus on serving classic French dishes
alongside innovative preparations highlighting California's wild game, fresh
seafood, and flavorful produce.[48]

Fior d'Italia

Angelo Del Monte came to San Francisco from Liguria in northern Italy. He
grew up helping with the family's boardinghouse in the busy port of Genoa.
When his Gold Rush ambitions fell through, he turned to what he knew:

MAY 1 1886

Fior d'Italia
RESTAURANT
SAN FRANCISCO

MENU

Veal Saute	.05	Tenderloin Steak	.30
Calfs Brains	.05	Veal Cutlets	.10
Risotto with Clams	.10	Porterhouse Steak for 2	.60
Veal Scaloppine	.15	Chicken Broiled	.20
Calfs Liver	.15	Chicken Saute	.25
Fritto Misto	.20	Squab Casserole	.40
Frog Legs	.40	Tortellini Bologna	.05

Special Dinner with Wine 35 ¢

Figure 5.6.   Fior d'Italia menu, 1886.

feeding a bustling port's workforce. In 1886 he opened a restaurant at 432 Broadway, Fior d'Italia, a San Francisco institution lasting more than 125 years. Its beginnings were hardscrabble, backed up against Telegraph Hill: "with its square bay windows and hideous fire escapes, it [was] not a beautiful blossom of architecture."[49] In the beginning, prices were low, and the restaurant's service was quick and efficient. The menu offered several dishes for just a nickel: sautéed veal, calf's brains, or a plate of tortellini Bolognese. Veal cutlets cost just a dime, as did clam risotto.[50]

In 1897 Angelo Del Monte merged his restaurant with the upscale Buon Gusto across the street, run by his brother Ferdinando. Fior d'Italia began to be known as a place for elaborate, celebratory dinners, not just a quick bite to eat. At the turn of the twentieth century, a dollar would still get one a full meal at Fior, with *antipasti*, salad, a fish course, spaghetti, roast chicken, dessert, and coffee. But one could also splurge on a $3.50 menu, if one had remembered to order it three days in advance. The expensive set menu included fancier dishes such as pâté de foie gras, prosciutto, *pompano en papillote*, *carciofi all'inferno* (steamed baby artichokes in a spicy tomato sauce), *capretto al forno* (a roast of young goat), and a lobster omelet. One could also try the fritto misto, "each delicately fried tiny roll of batter containing a different surprise—an artichoke heart, a piece of chicken liver, a bit of brains, or some other tidbit."[51] Later in the evening, revelers might stop by for a quick snack of sand dabs au gratin with a glass of sauterne.

Destroyed in the great 1906 earthquake and fire, Fior opened again in a makeshift structure. The Del Montes rebuilt their restaurant at 492 Broadway, with their old employee and current partner Armido Marianetti. Marianetti's three immigrant brothers also worked for the rebuilt Fior, as did his two sons, shelling peas and stringing beans after school. In October 1909 the owners announced proudly that they could seat three hundred people and offered the "finest Italian dishes in the City," including chicken sauté Italian style.[52] A decade later, the restaurant had expanded to seat seven hundred people, and had a dance floor and a cabaret theater as well. Diners appreciated the minestrone, ravioli, the *antipasti*, the "delicious Italian sauces and gravies," and the *zabaione* custard for dessert. An Italian writer set part of his semiautobiographical novel, *Un italiano in America*, in Fior d'Italia around 1914. The author praised the "real Italian cooking"—chicken cacciatore, veal scaloppini Marsala, and other specialties:

> tortelli home-style . . . cut out with glasses, filled with ricotta, and sealed with the tines of a fork . . . Little squares of thick cream mixed with rum that were lit with a match. They quickly produced a beautiful flame that burned yellow,

then blue and green. [The waiters] made espresso coffee right at the table with a portable burner.[53]

These crowds of diners and dancers were not all Italians:

> Plenty of good Americans, of Puritan or Cavalier stock for generations, like Italian cooking as well as anybody else. . . . The Fior d'Italia's pastries and cakes are joys to the gourmet, be he Italian, French, English or American.[54]

Under Prohibition, Fior found ways to continue to serve alcohol surreptitiously. Armido's son George Marianetti remembered sitting in a back room with a fifty-gallon barrel of homemade wine and a rubber hose for a siphon:

> The waiters would come by and say in a loud voice: "Six cupsa coffee." I would suck on the siphon, get the wine started. Fill six cups and slide them out. . . . I had to suck up a mouthful of wine each time. I was always a happy kid.[55]

During the Depression, the restaurant moved to a smaller space at 504 Broadway, the former home of the beloved restaurant Il Trovatore. Fior could now only serve two hundred customers at a time. Its dance floor and cabaret were gone. However, Finocchio's opened upstairs in 1936, filling the entertainment gap with a lively show of female impersonators.

After the war, Fior's reputation continued to rise under the guidance of the Marianetti family. Deanna Paoli Gumina, historian of San Francisco's Italian community, reminisced that Fior d'Italia had "the fanciest Italian meals" and was a favorite place to celebrate weddings, baptisms, first communions, and important birthdays.[56] In 1954 Fior moved to a corner of Washington Square, near the Italian church of Saints Peter and Paul. (Enrico's Sidewalk Cafe later took over the space at 504 Broadway.) San Francisco's politicians such as George Moscone and Joseph Alioto continued to treat Fior d'Italia almost as their lunchroom. Another former mayor, Dianne Feinstein, remembered lunches at Fior fondly: "they had the biggest, best, crispiest calamari."[57] Other popular dishes from the 1950s included cannelloni *alla toscana*; saltimbocca; beef *cenerentola* (a filet with artichoke bottoms stuffed with truffles); and *cima alla genovese* (a pocket of veal meat stuffed with spinach, pine nuts, and vegetables, served cold with other *antipasti*). These elaborate dishes led up to Fior's spectacular flambés, from fried cream to cherries jubilee and crêpes suzette.

Fior was a part of San Francisco and part of its evolving cuisine. Wholesalers such as Bruno Andrighetto remember the restaurant's changing expectations for produce: "George [Marianetti] wanted me to get him *radicetta*. . . .

Then they went into red lettuce. Then into romaine and then into the baby salads, just like everybody else because times change."[58] Even as the restaurant continued to serve spectacular flaming desserts, customers also appreciated the staff's focus on fresh, simple food. One such customer was Angelo Pellegrini, an Italian immigrant who helped inspire restaurateur and food activist Alice Waters, food writer Ruth Reichl, and the California Cuisine movement. In his 1956 book, *Americans by Choice*, Pellegrini praised Fior d'Italia for catering to the wholesome culinary tastes of Italian Americans:

> They feel perfectly at home in the restaurant they have molded to their . . . standard of culinary excellence. It is a very sane standard. Let there be cauldrons of soup, bushels of green and leaf vegetables, always cooked just so with the appropriate condiment. The olive oil must be pure, virgin, nutty in flavor. The meat, fish and fowl always fresh. Bring the salad to the table in half-barrel bowls, with an uncomplicated dressing of olive oil, wine vinegar, salt and pepper. . . . Good bread, good wine, genuine cheeses, ripe fruit, regional dishes, such as polenta or gnocchi at regular intervals, male waiters—and cleanliness.[59]

The insistence on male waiters jumps out because the rest of the paragraph sounds so fresh and modern. Fior, now located at 2237 Mason Street, has made a specialty of finessing that distinction: maintaining traditional forms and classic dishes while also keeping up with new trends in California Cuisine.

Schroeder's

In 1893 Henry and John Schroeder opened a saloon at 1346 Market. Their father had shipped over from Hanover in 1861, becoming a successful liquor importer and a prominent member of San Francisco's large German community. After he died, his sons split up. Both stayed in the liquor industry, however, until Prohibition made that impossible. Henry established a popular saloon at 545 Market, building up a steady clientele with his friendly manner and "the excellence of the viands he served." John bought and sold stakes in several different saloons and restaurants during the same period. After the earthquake, John opened a new saloon at 117 Front Street—but by 1910 he moved on, leaving the Front Street saloon for Henry. Henry stayed there until Prohibition, then moved to a smaller location just down the street. When he died in 1921, his widow sold the saloon and the name to Max Kniesche, who had been working at some of the best German restaurants in San Francisco and had a vision for how to keep Schroeder's going without the alcohol.

Kniesche was born in Prussia and arrived in San Francisco just after 1906. Over the next ten years, he worked at Tait's Cafe on Van Ness, at the

Heidelberg Inn, and at the Bismarck Cafe (he was there when H. L. Hirsch renamed it the Hof Brau). The Heidelberg's chef made a fantastic sauerbraten, a balance of savory and sweet. The Bismarck Cafe offered beef brisket, veal kidneys, and potato pancakes.[60] In 1916 Kniesche started working at Adolph Beth's Cafe across the street from the Hof Brau. Soon he and two other waiters bought the restaurant, renaming it the Nurnberg. They made money from alcohol and brought people in the door by offering a free lunch. "We had roast beef," he reminisced sixty years later: "We had ham and cold cuts. . . . It really was nice, that free lunch. It all disappeared afterwards."[61] Prohibition put an end to the free lunch. His partners wanted to try bootlegging, but Kniesche disagreed. When he left, he bought Schroeder's.

Kniesche brought in a favorite chef and promoted Schroeder's for its solid German merchant's lunch. They offered dishes such as bratwurst, sauerkraut, red cabbage, Wiener schnitzel, pot roast, meat dumplings, and their famous sauerbraten with potato pancakes. The menu was never exclusively German fare; Schroeder's also served paprika chicken, corned beef and cabbage, and, later on, lamb curry and rice. Kniesche did try to feature the many wines and beers produced by Germans in the Bay Area. During the 1939 Golden Gate International Exposition on Treasure Island, Kniesche built up Schroeder's reputation as one of San Francisco's culinary traditions. He had the artist Herman Richter paint timeless restaurant scenes for his walls, paying him only in food and drinks. In 1959 Schroeder's moved to 240 Front Street, where it still is today. In 1972, celebrating his fiftieth year at Schroeder's, Kniesche finally allowed women in at lunchtime.

Clinton's Cafeteria

At the 1893 World's Fair in Chicago, John Kruger introduced Americans to the word "cafeteria." The name meant coffee shop in Spanish, even though the model for Kruger's self-service concession stand was the Swedish smorgasbord. Five years later, Childs restaurant in New York introduced trays supported by rails to simplify self-service dining. In 1906 the Boos brothers opened a popular chain of cafeterias in Los Angeles, which was the first time many Californians experienced the concept. For San Franciscans rebuilding after the earthquake, cafeterias represented a new, modern way to eat.

In 1912 the first San Francisco Boos Bros. cafeteria began competing with E. J. Clinton's local chain of lunchrooms. Clinton quickly installed a cafeteria of his own, expanding on the Boos Bros. concept:

> He has followed the rule of giving people what they pay for, his principle being good service, good food, cooked by experts with a knowledge of hygienic laws

and sold at reasonable prices. . . . The ingenuity displayed in the utilization of space, the wonderful automatic conveyances, so that food is not touched by hands, the marvelous machines . . . peeling potatoes by the tubful at once, making piecrust for 100 pies simultaneously . . . everything working in harmony and co-operation, and under conditions of absolute cleanliness, strikes one with positive wonder at the achievement.[62]

Other cafeteria chains followed, including Compton's, which played a minor role in the 1934 general strike and then a major role in the August 1966 Tenderloin riot. But Clinton's cafeterias got off to a fast start. By 1921 Clinton's two downtown cafeterias together were serving more than five thousand people a day; moving that volume of food gave him access to the pick of the market in meat, seafood, and produce. A year later, a *Chronicle* article titled "Clinton's Pure Foods Explain Big Success" elaborated on the magnificence of these ingredients:

The tomatoes in their boxes all seem fresh and ripe; the peaches are clean and without spots; the apples have that fresh fragrance belonging to this year's crop; the lettuce is solid and crisp; the meats all smell fresh and pure.

One of the specialties of Clinton's is the unbeatable baked potato served with slices of butter and a dash of paprika. When you see them in stock, you know why they taste so delicious, for they are of the best Idaho variety, which Clinton buys by the carload.[63]

Women made up at least half of the clientele, which was very different from most San Francisco eateries. The new cafeterias seemed clean and modern. They did not serve alcohol, confirming the wholesome flavor of the institution. Fresh flowers were on the table and a small orchestra played from 6 to 8 p.m. every evening. For working women trying to get by on a small budget, cafeterias made sense.[64]

However, not all women loved the cafeterias. In 1921 Almira Bailey wrote:

Some people are very sensitive about cafeterias. They are cafeteria wise, they have a cafeteria class consciousness. . . . They have accurate minds which enable them to choose a well-balanced meal at minimum cost. . . . They don't take food just because it looks delicious. . . . They have a plan and stick to it. Wise and strong-minded, they shuffle their way bravely to the end. . . .

In California, [cafeterias] have attained a dignity and even lavishness that makes them the surprise and delight of the tourist. . . . We have music in our cafeterias. We have flowers on the tables. People don't just eat in them, they DINE. . . . We take visitors to see them. We brag about them, and when we wish to be especially smart we pronounce them caffa-tuh-REE-ah. Personally, I am proud of our cafeterias, but I do not get on in them.[65]

Bailey declared herself liable to take two salads, two desserts, or bread *and* a hot scone, helpless to resist. She teased her fellow San Franciscans for taking pride in their cafeterias, but that civic emotion built on a long history of pride in California's agriculture and food systems.

Trader Vic's

Victor Bergeron was probably the only San Francisco restaurateur to build his clientele on invented stories of his adventures with sharks and South Seas islanders. He was certainly the only one for whom a twenty-year-old Herb Caen created this backhanded compliment: "the best restaurant in San Francisco is in Oakland."[66]

In 1934 Prohibition had just ended, but the ongoing Depression meant people did not have much to celebrate. Bergeron took about $800 in family savings and opened Hinky Dink's, a small bar and sandwich shop at 6500 San Pablo Avenue in Oakland. His French-speaking immigrant parents ran a nearby grocery, so Bergeron knew the neighborhood well. He built up a solid clientele with his charm and his talent for cooking a savory meal out of almost anything. Bergeron had the imagination to see that Americans were eager to mark the end of Prohibition by trying fun new drinks, and so in 1936 he traveled to New Orleans and Havana, Cuba, in search of creative tropical drinks. Hinky Dink's ads promoted the results, boasting that "Vic" had brought back from Cuba "a full cargo of drinks and ideas direct from their creators."[67] Customers could try a Constantine Daiquiri (invented for Ernest Hemingway by Constantino Ribalaigua Vert at the Floridita bar in Havana); a Pino Frio (similar to a piña colada, without the coconut syrup); a Sazerac (with Peychaud's Bitters imported from New Orleans); or a Presidente (named for President Gerardo Machado, who ran Cuba from 1925 to 1933). Sometime in 1937 people began calling Bergeron "Trader Vic," and he played up his South Seas adventures even before he had any to mention.

Herb Caen's column spread the word about Trader Vic's, and customers drove across the new Bay Bridge to try his drinks. The 1939 Golden Gate International Exposition on nearby Treasure Island brought even more new customers to the restaurant and gave Trader Vic's national renown. Other bars and eateries started mining the same vein, combining tropical drinks and Tiki carvings, but they did not have Trader Vic with his genial warmth and wooden leg. (Anecdotes about shark attacks notwithstanding, Bergeron actually lost his leg to osteomyelitis as a child.) Nor did other restaurants have his flair for creative drinks and for culinary appropriation—he hired talented Chinese cooks to adapt Asian and island dishes to American tastes.

Figure 5.7. Trader Vic's restaurant. Photo Phil Fein. Courtesy of the San Francisco History Center, San Francisco Public Library.

In a 1958 *Newsweek* interview, Trader Vic declared: "The real, native South Seas food is lousy. You can't eat it." What he was offering was a fantasy, "complete escape."[68]

And it caught on. In 1941 *Harper's Bazaar* pronounced Trader Vic's "by far the most attractive and authentic of the many Polynesian places that have sprung up recently over America." In 1943 *Sunset* wrote about the restaurant's large Chinese barbecue ovens, burning white oak logs to produce an intense heat for Peking duck and other slow-cooked meats. In 1944 *Life* covered Trader Vic's, playing up the owner's adopted South Seas trader persona and his "resourcefulness with rum."[69]

The *Life* article lavished attention on the restaurant's outdoor kitchen, where "Chinese chefs barbecue and roast steaks, squab, ham, spareribs and even whole suckling pigs or wild boars." Photos included a barbecued pig surrounded by gardenias and ferns, a uniformed officer sharing a scorpion bowl with a friend and two young women, and Trader Vic relaxing in a Philippine cane chair with an amulet around his neck. The article praised Trader

Vic for sending ingredients for his favorite tropical drinks to U.S. troops in the South Pacific and the makings of hot buttered rum to U.S. fliers during the Aleutian campaign. His clientele in Oakland enjoyed an assortment of potent beverages in two-quart bowls: favorites included the Kava Bowl, the Scorpion Bowl, the Tiare Tahiti, and the B-17 Gremlin, named for the elusive gremlins who caused mechanical problems for World War II aviators. Trader Vic skillfully combined patriotism with exoticism, drawing an emotional bond from the home front to the troops fighting in the South Pacific.

Trader Vic's also benefited from a postwar fascination with the South Pacific, reflected in James Michener's novel *Tales of the South Pacific* and the Rodgers and Hammerstein musical *South Pacific*. In 1945 San Francisco's venerable Fairmont Hotel opened the Tonga Room & Hurricane Bar around the hotel's basement swimming pool. Polynesia had come to Nob Hill. Trader Vic's expanded to other locations, first Seattle and then Hawaii. Then, in 1951, the city of San Francisco got its own Trader Vic's, at 20 Cosmo Place, a few blocks from Union Square. This new location became the flagship of the growing chain; by 1953 Herb Caen wrote that he could no longer persuade San Franciscans that Trader Vic's started in Oakland. It was now a San Francisco institution, "with a branch across the Bay."[70]

The San Francisco Trader Vic's built a reputation beyond the Tiki trend with its "Captain's Cabin," a room where local politicians and VIPs socialized and ordered dishes that had not been on the menu for years. Caen continued to promote the restaurant, praising Bergeron's "passionate devotion to the art of haute cuisine" as well as his "marvelous" appetizers, from bacon-wrapped rumaki and cream-cheese-stuffed crab rangoon to lobster mousse and caviar sourced from the Columbia and Sacramento Rivers. The restaurant was also known for barbecued Indonesian lamb roast, curried chicken-in-a-coconut, garlic pake crab, Hong-Kong-style butterfly steak, and kidneys Martinique flambés. And everyone seemed to like Bongo Bongo soup, a spinach and oyster purée.[71]

In 1972 the original restaurant moved from the San Pablo location to Emeryville. In 1983 Queen Elizabeth and Prince Philip joined President and Mrs. Reagan for dinner at the San Francisco Trader Vic's. But the Tiki trend was on the way out, and in 1994 the prominent San Francisco location closed its doors. When the millennial generation sparked a nostalgic Tiki mini-craze, a new San Francisco Trader Vic's opened on Golden Gate Avenue near Van Ness, but it was only open for three years. The Emeryville restaurant is the only remaining Bay Area Trader Vic's. But there are many elsewhere: after Vic introduced San Franciscans to his vision of South Seas cuisine, he spread that vision around the world.

India House

In 1947 the British ceded formal imperial control over India. That same year, the British immigrant David Brown gave up working for Shell Oil and opened the India House restaurant, bringing a small taste of colonial nostalgia to San Francisco. The first incarnation was on Clay Street and had just ten tables and one waiter. By 1949 India House had moved around the corner to a larger space on Washington. Brown hired Indian and Pakistani students to serve the food and add color with their "native coats and turbans."[72]

At first, India House did not serve cocktails, but the clientele began asking for them. Soon people came as much for the colonial atmosphere of the bar as for the food itself. They drank sherry, Pimm's Cup, or gimlets, while noshing on tiny curried meatballs or bubbling hot deviled shrimp served in half clam shells with adorable little forks. Once seated, diners ordered curries made with chicken, beef, lamb, meatballs, clam, or prawns. An assortment of condiments arrived on the table:

> Grated coconut, Major Grey's mango chutney, sweet pickle relish, chopped egg, ground peanuts, raisins, quarters of fresh limes (particularly good with the seafood), and fresh onions . . . as well as hotter sauces for the incurably adventuresome.[73]

The restaurant made chapattis and poppadoms, samosas and dal (lentil stew), and colonial soups such as mulligatawny and Senegalese, the latter made with chicken, curry, and heavy cream, and then served cold. Dessert might be sliced mangoes, rice pudding, or *gulab jamun* (a cheesy fritter).

In 1968 three former servers at India House bought the restaurant from David Brown. South Asian immigrants, the new owners nevertheless kept much of the colonial atmosphere, including the popular Pimm's Cup. They kept the loyalty of their old patrons—but they also attracted new customers by bringing in a tandoor oven. Sarwan Singh Gill, one of these new owners, came from India's northern Punjab region. Tandoor cooking from the Punjab was a new trend in Indian food, as people in San Francisco and around the world fell in love with tandoori grilled meats and butter chicken.[74] Gill and his co-owners moved the restaurant to 350 Jackson in 1972. India House remained popular for decades, and only closed in 1995.

Enrico's Sidewalk Café

Before the mid-1950s, North Beach was not known for its café scene. But a new generation of Italian Americans had more free time and more cash to

Figure 5.8. Lunch crowd at Enrico's restaurant. Courtesy of the San Francisco History Center, San Francisco Public Library.

spend than their parents, and the rise of the Beats gave San Francisco coffeehouses a fashionable allure. At the time, Enrico Banducci (who was born with the first name Henry but preferred to go by Enrico) was already running a famous nightclub, the hungry i, on the ground floor of the International Hotel. In 1958 he saw the success of the new Caffè Trieste and opened his own coffeehouse, Enrico's Sidewalk Café, at 504 Broadway.[75]

Enrico's was a great place to see and be seen, with tables on a patio across from the City Lights bookstore. Every night one could sit outside at Enrico's and watch the crowds on Broadway and the line to get into the famous drag shows at Finocchio's upstairs. The restaurant was known as a place to people watch rather than for its food. But diners came back for Enrico's trio of mini-hamburgers with cornichons, as well as creative salads and dressed-up tavern food such as Welsh rarebit with bacon and grilled tomato. In later years, the café also served a California bistro menu: smoked salmon bruschetta, Niman Ranch flatiron steak with lyonnaise potatoes, and a popular thin crust pizza, along with San Francisco's first mojitos.

For most of its life, Enrico's featured a cast of memorable characters. There was Banducci himself, always at the café wearing his trademark beret and treating his friends to drinks or tasty slices of baklava. Ward Dunham was behind the bar, mixing drinks for Enrico's for more than twenty years. Famous regulars included Herb Caen, Willie Brown, Ron Kovic, Francis Ford Coppola, and radio personality Scott Beach. Poet and novelist Richard Brautigan spent many hours at Enrico's, making a habit of welcoming newcomers to the city and to North Beach.

The Mandarin

Cecilia Chiang, founder of The Mandarin, grew up in Beijing in a wealthy household. Her mother had both a local cook and another cook who had come from Shanghai with the family when they moved north. Chiang and her many sisters were not allowed in the kitchen with the male cooks, but they learned the nuances of good Chinese food by eating well. During the war, Chiang escaped Japanese-occupied Beijing and traveled on foot through Anhui Province, Xian, Sichuan Province, Chongqing, and on to Shanghai. When the Communists took over in 1949, she fled with her husband and children to Tokyo, where she opened a Chinese restaurant called The Forbidden City. She still did not know how to cook, but she hired professional chefs and used her diplomatic connections to get decent ingredients.[76]

In 1960 Chiang came to San Francisco to visit her sister. She helped some other Chinese immigrants with what was supposed to be a short-term loan, but when they walked away from their proposed restaurant, she found herself suddenly opening a restaurant in a city where she did not speak either of the two main languages, English or Cantonese. Given her Tokyo experiences, however, she decided to forge ahead, introducing San Francisco to a new kind of Chinese restaurant. She respected restaurateurs such as Johnny Kan who provided excellent Cantonese food, but Chiang opened The Mandarin to serve northern Chinese cuisine.[77]

She hired a couple from the northern Shandong province to cook, and for the front of the house she hired Berkeley students from very well-educated Chinese families, who spoke excellent English. Instead of chop suey and egg foo yung, she served spicy Sichuan-style dry-shredded beef, slender Chinese eggplant, tea-smoked duck, beggar's chicken, handmade pot stickers, and sizzling rice soup. As Chiang recalled in 2006: "Soup poured into sizzling rice. Zsa! Dramatic!"[78] Her expenses were high because she had to import specialty ingredients from Taiwan, but she understood she was making a real contribution to her adopted city. San Francisco had never seen these kinds of dishes.

Chiang used her warmth to establish her business, becoming friends with Trader Vic and Alexis Merab, a Russian restaurateur with ties to Shanghai. Through their influence, Herb Caen came to The Mandarin, enjoyed himself, and raved in his column about her "hole-in-the-wall" restaurant on Polk. The Mandarin gained a solid reputation for its food and elegant service, and in 1968 Chiang moved the restaurant to a larger space in Ghirardelli Square, where it thrived. Her basic rule for decorating The Mandarin: "No gold. No red. No dragons. No lanterns." Starting in the 1970s, Chiang was part of the new wave of California restaurateurs, promoting fresh, seasonal cooking with an ever-changing selection of local meats, seafood, and vegetables.[79]

Chez Panisse

One of San Francisco's most famous restaurants is not even in San Francisco. In the late 1960s Alice Waters was living in Berkeley but missing France and cooking meals based on Elizabeth David's books. She learned from David to love fresh market ingredients cooked with simplicity. Over time, the new California Cuisine movement would emerge out of that approach. In 1968 the alternative newspaper *San Francisco Express Times* gave Waters the opportunity to promote her favorite recipes in a column she called "Alice's Restaurant"—a play on Arlo Guthrie's 1967 antiwar song. The young Waters featured comfort foods such as chocolate mousse, applesauce, and cheese and onion pie, as well as more challenging dishes such as pâté, borscht, "orange duck," and chicken biryani.

Her friends encouraged her to take the next step and open a real restaurant. In 1971, with their help, she opened Chez Panisse, named after Waters's favorite French film character. Alice and her friends had never run a restaurant before, and they served ambitious food they had no experience cooking. From the day it opened, Chez Panisse was like no other restaurant, and crowds came to experience Waters's vision. But the restaurant swallowed up its investors' money and did not turn a profit for more than a decade. Part of the problem was that the staff drank the restaurant's good wines after closing. Another problem was that Waters's own ethos and her network of countercultural Berkeley suppliers precluded raising her prices as high as the market would bear.[80]

Waters hired a series of chefs, building menus with them to link her vision of French country cooking with their particular talents. Although some dishes returned year after year, the restaurant constantly pushed to break new ground with brilliant new dishes and themes for its set menus. In early 1973 Waters hired Jeremiah Tower whose unapologetic East Coast authoritarianism and classical haute cuisine training shook up the Chez Panisse kitchen. At first, he cooked Waters's simple bistro selections, from quiche to bœuf bourguignon.[81] But his ambition drove him toward elaborate dishes such as sweetbreads cooked in brioche pastry with a champagne sauce. He also explored the Bay Area for the finest ingredients, from ducks and Monterey Bay prawns to wild boar and fresh organ meat. While Tower focused on meat, Waters worked to expand the range of salad greens she could offer, persuading friends and local farmers to plant sorrel, radicchio, or mâche from seeds she had sourced from France. And everyone on the staff looked out for cheap but succulent seafood.

In search of greater fame, Tower created and publicized one-of-a-kind dinners. He reached out to influential writers such as Herb Caen and James Beard, with meals billed as "the Alice B. Toklas Cookbook dinner," or a Brittany-themed dinner featuring crêpes de moules, canard nantaise (roast duckling with baby peas), watercress salad, Pont L'Évêque cheese, and an almond cake. In February 1975 the restaurant threw a Curnonsky festival, serving beef marrow tart, braised duck stuffed with shallots, and other dishes based on the teachings of the famous French gastronome:

Cooking is when things taste like themselves.
In cooking as in all the arts, simplicity is the mark of perfection.

The adventures in dining that Waters and Tower cooked up echoed their experiences in French bistros, except that they were inventing new dishes every night without any training or infrastructure.[82]

In October that year *Gourmet* published a glowing review, praising the chef's inventiveness within the French tradition. Tower responded with an outlandish Salvador Dali dinner featuring a leg of lamb injected with Madeira and brandy, or, as the menu had it: "drugged and sodomized." On July 14, 1976, Chez Panisse threw the first of its annual Bastille Day garlic extravaganzas, celebrating the French national holiday with very un-French meals. The set menu featured baked garlic and mushrooms in grape leaves; beef tripe with basil and garlic; fresh figs served with white cheese and garlic honey; and six other garlic-infused courses. That fall, Tower dreamed up a "Northern California Regional Dinner," exalting the region's offerings with dishes such as Big Sur Garrapata Creek smoked trout steamed over California bay leaves. The ingredients were all local, including Tomales Bay oysters, Sebastopol geese, and a Schramsberg Cuvée de Gamay rosé. Even as it echoed

CHEZ PANISSE
NORTHERN CALIFORNIA REGIONAL DINNER
OCTOBER 7, 1976

Spenger's Tomales Bay bluepoint oysters on ice

Cream of fresh corn soup, Mendocino style, with crayfish butter

Big Sur Garrapata Creek smoked trout steamed over California bay leaves

Monterey Bay prawns sautéed with garlic, parsley, and butter

Preserved California-grown geese from Sebastopol

Vella dry Monterey Jack cheese from Sonoma

Fresh caramelized figs

Walnuts, almonds, and mountain pears from the San Francisco Farmers' Market

$20.00

Figure 5.9.   Chez Panisse menu, 1976.

a grand October 1895 California-themed banquet at San Francisco's California Hotel, Tower's dinner pointed the path forward, underscoring Chez Panisse's growing fame for simple dishes made with fresh, local ingredients.[83]

In 1977 Tower left the restaurant, and Waters remained to realize that vision of a simpler cuisine, focused on the neighborhood rather than the international culinary scene. She began strengthening the network of providers on whom she could rely. Tower's well-trained sous-chef, Jean-Pierre Moullé, took over in his place and over time demonstrated that he shared Waters's mission. Many people cooked at Chez Panisse over the years, alongside Moullé or at lunch. Others worked in the less formal upstairs café, making simple dishes from the same fresh ingredients as the restaurant downstairs. Customers relished pizzas from the café's wood-burning pizza oven, inspired by a pizzeria in Turin, Italy. To this day, a reservation downstairs at Chez Panisse is a treasured prize.

A surprising number of Chez Panisse alumni went on to start influential restaurants of their own around the Bay Area or further afield. Jeremiah Tower founded the eclectic restaurant Stars in a prime location near City Hall; Judy Rodgers opened Zuni Café on Market Street, where she was joined by another member of the Chez Panisse family, Gilbert Pilgram; Deborah

Madison founded Greens in the Fort Mason Center, serving elegant, simple vegetarian cuisine; Joyce Goldstein started Square One in the Financial District; Mark Miller opened the Fourth Street Grill and the Santa Fe Bar and Grill, both in Berkeley; Christopher Lee started Eccolo in Berkeley as well; and Paul Bertolli opened Oliveto's in Oakland.

Chez Panisse was also influential in encouraging specialty shops and farmers who shared Waters's approach. In the early days, Chez Panisse staff would pick up good deals on delicious offerings from the Cheese Board across the street, or wines from Kermit Lynch. Later, restaurant staff went on to start Acme Bread, the Cowgirl Creamery, Pig-By-The-Tail Charcuterie, and the Monterey Fish Market. Waters also gave strong support to innovative goat cheese producer Laura Chenel, organic dairy pioneer Ellen Straus, rancher Bill Niman, and local farmers Bob Cannard and Joe Schirmer, among others. Sibella Kraus, a former chef and forager for Chez Panisse, went on to found several San Francisco food institutions: the groundbreaking Farm-Restaurant Project, connecting Bay Area chefs with local organic farmers; the Center for Urban Education about Sustainable Agriculture (CUESA); and the San Francisco Ferry Plaza Farmers' Market.

Chuck Williams's cookware company predated Chez Panisse by fifteen years, and by 1958 Williams Sonoma had a flagship store on 576 Sutter Street. But the new California food movement of the 1970s sparked the company's rapid growth. California Cuisine emphasized fresh, local ingredients with simple preparations to bring out their flavors. Chez Panisse showed how spectacular the resulting dishes could be, and Williams Sonoma offered untrained cooks the equipment to make similar dishes at home.[84] Chez Panisse and Williams Sonoma complemented each other well, both institutions building off the work of Elizabeth David not only to change what kinds of food people enjoyed, but also to inspire a different kind of home cooking, both simple and satisfying.

CHAPTER SIX

~

San Francisco Cookbooks

From very early on, San Franciscans published cookbooks to promote a Northern California approach to cooking. Chefs realized that many people were interested in San Francisco cuisine. Cookbooks reached out to those who could not travel regularly to enjoy the city's restaurants and also to those, near or far, who appreciated San Francisco's cosmopolitan cooking and hoped to re-create it at home. Different cookbooks represented different ethnic approaches to cooking in California, but a common thread emphasizing fresh ingredients runs through most of them. The two earliest San Francisco cookbooks were published in 1872. One was titled *How to Keep a Husband or Culinary Tactics* and the other, somewhat less poetically, the *California Recipe Book*.

California Recipe Book (1872)

The *California Recipe Book* started with an ad for Haynes & Lawton on Market Street, importers of fine French china, and ended with several more pages of ads, revealing straightforwardly the book's role in a culture of conspicuous consumption. The book's first words of advice recommended obtaining the best possible flour, "even at greater cost." The exhortation to spend as much as necessary to obtain the best ingredients would become a common piece of advice in Northern California. The book's recipes followed each other in rapid fire, each credited with an initial and an address: Mrs. C. on Harrison

Street, Mrs. O. on Bush Street, Mrs. E. on Mason Street, and so forth. They included measurements but assumed the reader knew how long to bake muffins or how to make a basic pie. For instance, Mrs. C. on Ellis Street provided only the following:

> **Whortleberry Pie:** Fill the dish not quite even full, and to each pie of the size of a soup-plate, add four tablespoons of sugar. Dredge a little flour over the fruit before you lay on the upper crust.[1]

Similarly, Mrs. B. on California Street gave a succinct recipe for doughnuts: "1 cup sugar, 1 cup milk, 1 egg, 1 tablespoon melted butter. Allspice for flavoring." The rest was up to the reader. The cookbook offered mostly recipes for baked goods, preserves, and puddings; several of the puddings and flummeries used packaged gelatin, but most did not. Mrs. R. provided a section on cooking fish, from a fish chowder that starts by frying up four or five slices of fat salt pork, to her advice on striped bass: "Disliked by many, as usually cooked, would be more generally approved if cooked long enough." Another section, copied from an earlier Boston cookbook, explained how to prepare both green and black tea, as well as how to make a breakfast drink from boiled cocoa shells.

Some dishes sound quite odd to modern readers: Ground Rice Pie called for soaking rice in milk, boiling the rice a few minutes in water, then mixing in eggs, and baking in a pie crust. Others offered common ways to feed a family on a budget: "summer mince pies, without meat" involved mostly crackers, molasses, vinegar, sugar, and chopped raisins, along with "spices without stint." It is pleasant to picture these women having easy access to spices when meat was out of reach. Beef was mostly absent from the cookbook, although one recipe explained what to do with a calf's head (start by tying up the brains in muslin, together with sweet herbs). Included were three recipes for chicken salad, and one each for deviled crab, scalloped oysters, fried oysters, and terrapin. The book did not cover much in the way of vegetables: besides corn and a bit of lettuce and celery in the chicken salads, it addressed only tomato catsup, preserved green tomatoes, pickled green tomatoes, and chili sauce.

The tone varied by contributor. Some recipes were cheerful and suggested the results would be "splendid," whereas others were more modest and simply gave the instructions. The unnamed editors provided a few options for readers—noting that "pie-crust looks nicer made of lard; but tastes better half butter"—but for the most part they stayed out of the way of the recipes, allowing the different women to speak for themselves.

How to Keep a Husband or Culinary Tactics (1872)

Like the *California Recipe Book*, the compilers of *How to Keep a Husband* included numerous advertisements to help bring in money. It was probably published as a community fundraising project for an unnamed church, to judge from the advertisements promoting Bibles and prayer books. Right next to those ads were others for sewing machines, fine china and silverware, butchers, grocers, and shellfish mongers, and even the very latest in kitchen technology—the iron cookstove. Apparently the cookbook committee did not object to the juxtaposition of religion and commercialism. The committee also embraced the idea that a woman should cook strategically to please her husband, rather than herself or her children. Ads promoted "Ladies' French Underclothing" as well as a "Ladies' and Gentlemen's Oyster Room," which was open at the California Market from 6 a.m. until midnight every day.[2]

The compilers of these contributed recipes set the bar high for the level of cooking knowledge they expected in their community. The meat section, for instance, started off dismissively: "Directions for plain Roast and Boiled meats can be found in every Cook-book; to these we refer the ladies." In their book, they intended to provide recipes not just to feed a family, but to win a husband's love again every night. That impulse seemed to lie behind numerous recipes, including this firm instruction:

> All roasted game, birds or chickens are improved by larding them first with tiny strips of salt pork, drawn through the skin with a larding needle, each stitch an inch long, leaving an end of pork half an inch out. This dresses the roast, and looks well when brown.

The first pudding recipe was an old poem suggesting that the point of pudding was to please one's husband:

> Eve's Pudding:
> If you want a good pudding mind what you are taught
> Take of eggs six in number when bought for a groat
> The fruit with which Eve her husband did cozen,
> Well pared and well chopped, at least half a dozen;
> Six ounces of bread, let Moll eat the crust,
> And crumble the rest as fine as the dust.
> Six ounces of currants from the stalks you must sort,
> Lest you break out your teeth, and spoil all the sport;
> Six ounces of sugar won't make it too sweet,
> Some salt and some nutmeg will make it complete;

> Three hours let it boil without any flutter,
> But Adam won't like it without wine and butter.

Molly, the generic Irish cook, got only dry crusts, while one's husband ate pudding, made with wine and butter to make it even richer. The cookbook's more concise recipe for "Irish pancakes" reflected a similar approach to pleasing one's husband by combining sweet and starchy with alcohol: "Twenty-four eggs, two quarts milk, one-half pint of beer or three-quarters, one pound of sugar, two pounds of flour, ginger to your taste, a little spirits." Cherry Bounce was another recipe that came highly recommended by the book's recipe committee. By mixing wild, bitter black cherries with Jamaican rum, brandy, and a sugar syrup, one could produce "a delicious cordial; an excellent tonic," well suited to bringing the pep back into married life. The cookbook also offered up a sweet and spicy chili sauce, which went beyond previous English-language recipes for chili vinegar or spicy tomato sauce. This one added brown sugar to sweeten the sauce:

> **Chili sauce**: skin twelve large ripe tomatoes, chop them with four Chili peppers and two onions, add four cups of white wine vinegar, two tablespoons of brown sugar, one tablespoon of salt. Boil hard one and a half hours. Put up in wide-mouth bottles.

The editors provided many treats to please even the pickiest husbands, including a reliable recipe for strawberry shortcake:

> **Strawberry short cake**: Rub one teaspoon of cream of tartar into one pint of flour, a piece of butter the size of an egg, half a teaspoon of soda; mix it with cold water, and bake on a tin sheet. Split open while hot, butter and spread thick with berries, sprinkle with sugar, and cover with the other half of the cake; make a dressing of one tea-cup of strawberries, one tea-cup of boiling water, three large spoons of sugar, boil a few minutes, pour over the cake, put into the oven for a few minutes.

What Mrs. Fisher Knows About Old Southern Cooking (1881)

Almost a decade after those first volumes, the African American cook Abby Fisher put together an elegant cookbook, giving it a single voice that contrasts vividly with the multi-vocal nature of the earlier community cookbooks. In her preface, she apologized for her illiteracy but promised, with the help of her friends, to provide instructions for:

Soups, Gumbos, Terrapin Stews, Meat Stews, Baked and Roast Meats, Pastries, Pies and Biscuits, making Jellies, Pickles, Sauces, Ice-Creams and Jams, preserving Fruits, etc. . . . so that a child can understand it and learn the art of cooking.[3]

She mentioned nine friends who supported her cookbook project; some of them transcribed her recipes because she could not write them herself, while others helped cover the initial publication costs. She also proudly asserted that she was "Late of Mobile, Ala."—her origins gave her authority in southern cookery, even as she had demonstrated her confidence and cosmopolitanism by moving away from the South and making a name for herself in San Francisco.

Fisher included about the same number of recipes as the *California Recipe Book*, but in many more different categories: thirteen in all. She offered six kinds of salad, eight kinds of croquettes, twelve kinds of roast meat, fourteen kinds of soup, and twenty kinds of pickled vegetables—her particular culinary specialty. Her recipes were far more detailed and professional than the earlier San Francisco cookbooks. Compare the following recipes for ginger cookies:

Ginger snaps: ½ cup of brown sugar, ²/₃ cup of molasses, not quite ½ cup of water, ²/₃ cup of shortening, 1 teaspoonful saleratus, dissolved in the water, cinnamon, a very little ginger. (*California Recipe Book*)

Ginger cookies: One teacup of molasses, one-half teacup of sugar, one tablespoonful of butter, one tablespoonful of lard, one quart of flour, two tablespoonfuls of ginger, one teaspoonful of cinnamon, one teaspoonful of allspice, two tablespoonfuls of yeast powder. Cream butter and sugar together and add molasses. Sift yeast powder and flour together and add to butter, sugar and molasses, then add lard and spices, etc., and work it up well. Roll out on a board, and cut them out and bake like you would a biscuit. (*What Mrs. Fisher Knows*)

Fisher's recipe provided much more information, even if she saved space by referring to steps explained in other recipes. She mostly avoided recent commercial products, aside from yeast powder; she did use boxed gelatin for one pudding recipe and flavoring extracts for a cake. Her professional experience cooking for a wide range of customers shone through in the alternatives she offered in many recipes, advising readers to substitute thyme if they did not like onions in a stuffing, to add sweet cream to the terrapin stew "if you like it," or reminding them they could add more cayenne pepper to the chow-chow relish recipe "if you like it hot." She was also not shy about recommending her favorite dishes to readers: "You will find the calf's head soup the most delicious soup in the cookery. Study the recipe and remember it well." She called Spiced Round of Beef "the most delicious appetizer among meats." Another of her favorite recipes took a whole day to prepare:

> **Game Sauce**: Take one peck of plums, half dozen silver skin onions and chop them very fine; put on the plums to cook. First seed plums; use a porcelain kettle; put the onions to stew in a pint of vinegar until thoroughly done, then add them to the plums; four pounds of granulated sugar to be added; season with one teaspoonful of cayenne pepper, one of black pepper, two ounces of cinnamon broke in fine pieces; cook on a slow fire, stir frequently to avoid burning—one teaspoonful of table salt—it will take one whole day to cook; when cool cork in a tight jar and keep in cool closet.

In her informed opinion, this was "the best sauce in the world."

In her afterword to the 1995 reissue of Fisher's cookbook, historian Karen Hess provided an excellent guide to Fisher's life and times, offering tips for actually cooking from the book. She pointed out that Fisher would have learned to bake with a brick oven and to roast over a hearth but adapted her recipes to work with the iron cookstove, which was increasingly common in middle-class kitchens. Hess also noted a few mistakes: the 1881 transcriber wrote down "circuit hash" when Fisher must have said succotash; "Carolas" instead of crullers; and "jumberlie" instead of jambalaya. Andrew Warnes noted how the term "jumberlie . . . wonderfully retains [Fisher's] South Carolinian lilt," revealing her roots. This was a moment when Fisher's culinary expertise and cultural background overwhelmed her friends' efforts to pin down her recipes in standard written English.[4]

Clayton's Quaker Cook-Book (1883)

H. J. Clayton was another San Francisco character—a pioneer from the earliest days, he first ran "Clayton's Saloon" on Commercial Street, where "the most respectable citizens" would go to enjoy a broiled quail or oyster stew. In 1851 he was elected San Francisco's inspector of meats. In 1853 he opened the Clayton House, at the corner of California and Montgomery, where he promised customers a constant supply of fresh oysters. By 1866 he was running an Oyster and Chop House on the corner of Clay and Leidesdorff Streets and billed himself in advertisements as "the Prince of Caterers." In 1881 he began publicizing the imminent publication of his ultimate achievement, "a cook-book which was the quintessence of all knowledge on the subject."[5]

Rumors spread that Clayton was a woman who dressed as a man, and Clayton did little to contradict those stories. The frontispiece photograph of his cookbook was ambiguous in gender, and the book was printed at the Women's Co-Operative Printing Office in San Francisco. Writing in the third person, Clayton mentioned in his introduction that he had been raised on a farm, but "being in his younger days of a delicate constitution," he did not do much heavy farmwork. Instead, he stayed with his mother to help with "the culinary labors of the household."[6] For his 1881 lecture, he had a friend read his speech while Clayton prepared a tasting for the audience: clattering dishes around, occasionally "cocking his ears with a comical expression of amiable curiosity which made the audience roar with laughter, and then going on demurely with his task."[7]

In his cookbook, Clayton continued the performance of his well-honed character. In the section on pork, he went on at length:

> Corn, or any kind of grain-fed, or, more especially, milk-fed pork, as every one knows who is not of the Hebrew faith, which entirely ignores this—when properly prepared, well-flavored, oleaginous production—and is fond of pork, from the succulent sucking pig, the toothsome and fresh spare-rib, unrivalled as a broil, to the broiled or boiled ham, and side-meat bacon of the full-grown porker, is vastly superior to the meat of the slop and garbage-fed animal raised and slaughtered in the city—more especially as the butchering of hogs in San Francisco is at this time entirely monopolized by the Chinese population, who seem to have a warm side, in fact a most devoted affection, for the hog, surpassing even that of the bog-trotters of the "Ould Sod" for the traditional pet-pig that "ates, drinks and slapes wid the ould man, the ould woman, and the childer."

Once this former meat inspector finally got around to explaining how to select fine pork meat—by going to a reputable dealer when the lots arrive from the countryside in the fall and selecting a carcass of no more than a hundred pounds, with hard, white flesh and thin skin—it was almost beside the point.

If Clayton's tone in the above passage seems xenophobic, his cooking was less so. He made gumbo and Spanish omelet; he used curry powder in his split pea soup and chili sauce to dress cucumbers. His recipe for "Squash and Corn, Spanish Style" may be the first printed Mexican recipe in a California cookbook.[8] His recipes themselves were chatty and accommodating. When explaining how to make "Fresh Oyster Soup," he began by recommending thirty small Eastern and fifty California oysters, but the end of the recipe noted that for "a larger quantity of soup, add a can of good oysters, as they will change the flavor but little." Indeed, despite his early association with the city's oyster industry, Clayton did not shy away from recommending the use of "can oysters," along with their "rich gravy," in dishes such as oyster patties and oyster omelet. He also recommended the use of canned lobster, canned corn, and canned tomatoes, as in Clayton's Popular Sandwich Paste (made with slow-cooked ham, calves' tongue, canned or fresh tomatoes, mustard, and Worcestershire sauce), or his equally splendid Monmouth Sauce:

> **Monmouth Sauce:** In making this delightful ketchup, take 25 pounds of fresh, or two 8 lb. cans of tomatoes, and slice, not too thin, adding five medium sized onions cut fine. Put these, with plenty of salt, in a porcelain kettle; adding, with a handful of hot green peppers, or a less quantity, if dried, 1 ounce of white ginger, chopped fine, 1 ounce of horse-radish, and 1/2 ounce each of ground cloves and allspice, and 1 lemon, with seeds removed and cut small. After letting these boil for three hours, work through a sieve and return to the kettle along with a pint of wine vinegar, 2 tablespoonfuls sugar, 2 of good mustard, a teacupful of Challenge or Worcestershire Sauce, and let boil for 2 or 3 minutes, and set off. To prevent fermentation, stir in a teacupful of high-proof California brandy.

Historian Dan Strehl has described Clayton as a progenitor of the modern movement to eat locally and know the source of your food. Indeed, the author of the *Quaker Cook-Book* was a philosopher as well as a chef, advocating a Quaker-like simplicity in cooking, as well as high-quality ingredients. Clayton eagerly recommended particular shops and dealers in the Bay Area: Engelberg's bakery on Kearny, Davidson's cheese factory in Gilroy, Hills Bros. Coffee on Fourth Street, the Jersey Farm Dairy in San Bruno, and, for fresh meat, John McMenomy at the California Market. He also mentioned John Bayle, a butcher specializing in tripe, calves' heads, feet, tongues, ox tails, sweetbread, and brains, who had a market stall right next to McMenomy. As Clayton wrote in his introduction: "In these degenerate days of wholesale adulteration of almost every article of food and drink, it is eminently just and proper that the public should be advised where the genuine is to be procured." Despite his affection for particular dealers and his openness to using canned goods, he still felt some products needed to be made by hand. Thus Clayton explained how to make "nudels" for soup by hand without mentioning that Italian grocers had been retailing boxed noodles in San Francisco for twenty-five years.

El cocinero español (1898)

In 1898 Encarnación Pinedo published her own cookbook, the first written by a Hispanic within the United States. In 2003 Dan Strehl translated selections from that book and republished them in *Encarnación's Kitchen: Mexican Recipes From Nineteenth-Century California*.[9] Strehl's edition includes two excellent essays about the author and her place in history, by Strehl and journalist Victor Valle. In contrast to the illiterate Abby Fisher, Pinedo could read and write in both English and Spanish, and came from an important family of Californios, the Berreyesas, whose vast properties were seized by well-connected Anglos in the 1840s and 1850s. Numerous members of the Berreyesa family were killed by Anglos during the same period, without legal justification. Pinedo herself was born in 1848. As she grew up, she lived with her widowed mother and later with her elder sister, who had married an Anglo "omnibus" driver. She began to transcribe family recipes as a way to pass them on to future generations; her cookbook is dedicated to her nieces, so they would remember "the value of a woman's work." She did not, however, think culinary knowledge was only for the domestic sphere, or only for women who could not afford servants. She herself hoped to earn money from her aptitude in the kitchen. Rather than just keeping a personal notebook with her recipes, she sought a printer and advance sales for her cookbook. She had reason to think her culinary skills would find an appreciative audience, at least among the many who had enjoyed the Pinedos' generous hospitality for decades. (The Pinedos had a large house and were often called upon to host elaborate wakes in the Spanish Californian community, which could mean feeding a large crowd for two days.)

Her cookbook presented itself as a collection of family recipes with links to the old country; that is, to Spain. But she had to re-create the recipes based on her childhood memories and cookbooks she acquired as an adult, so they are rough approximations of what was cooked in Californio households during Gold Rush days. Likewise, one can question her assertion that these recipes have strong Spanish roots. In the 1890s race was increasingly the means by which people were categorized: Pinedo made a strategic choice to emphasize the "whiter" part of her heritage rather than her Mexican ancestors. The names of her dishes often referred to Spanish cultural markers such as *gallego* or *castellano* (Galician or Castilian), whereas the ingredients and techniques were more Mexican. Her recipes relied heavily on peppers, as well as fresh fruits, vegetables, and herbs; her spicing was more Mexican than Spanish in style. She also did not acknowledge her many debts to an older Mexican cookbook, the *Nuevo cocinero mexicano*.

In the cookbook, she did not dwell on the wrongs done to her family, although she let her feelings show in some places, as when she indicated her disdain for English cooking: "There is not a single Englishman who can cook, as their foods and style of seasoning are the most insipid and tasteless that one can imagine." Given the English roots of most of the recipes published in nineteenth-century Anglo cookbooks, that can be read as both an assertion of cultural pride and a dig at competing cookbooks. That said, she included some recipes from the English tradition, such as English-style rolls, English-style potatoes, and ham and eggs. She called ham and eggs "huevos hipócritas," although it appears that she took both the name and the recipe from the *Nuevo cocinero mexicano*, so any sting to the term "hipócritas" did not originate with her.

She included several Argentinean recipes, including an elaborate recipe for "Pasteles o empanadas a la argentina" (Argentine-style pastries or turnovers), stuffed with both beef and chicken and flavored with not just garlic, onion, salt, and pepper, but also tomatoes, green chilis, oregano, cumin, olives, and raisins. Her vast assortment of meatball recipes included not only German-style and Italian-style meatballs, but also "Albóndigas españolas a la alemana" (Spanish meatballs in the German style)—almost a dumpling. She told her reader how to make "Sandwiches de pâté de foie gras," preparing the bread with butter and a little French mustard. She also recommended "Chiles rellenos con sardina francesa en cajas" (peppers stuffed with canned French sardines) and approved of using imported French olives over local California ones. The mixing of traditions, even in a book she titled "The Spanish Cook," suggests that it was common in her community to enjoy favorite dishes from different cuisines. Pinedo's recipes represented the best of Mexico's colonial cuisine, combining cosmopolitan influences to produce a series of sophisticated dishes. Pinedo took freely from old and new culinary approaches, including well-established beefsteak and barbecue recipes, and a dish that she almost certainly ate at the mission as a child: *carne asada en la olla de los misioneros*. She also explained how to make *Tamales al vapor*, calling these steamed tamales the "ultima novedad," the latest thing. She was eager to spread the word about the new Enterprise Nixtamal corn mill, which she considered a great replacement for the metate in grinding a family's daily corn. Writing at the cusp of the twentieth century, Pinedo had a flair for updating the old ways and making new techniques seem familiar and accessible to her readers.

101 Epicurean Thrills

Starting in 1901, May Southworth put together a series of slim cookbooks for San Francisco publisher Paul Elder, each containing one hundred and one recipes. The first book in the series was *101 Sandwiches*, followed by *101 Entrées, 101 Beverages, 101 Salads,* and *101 Chafing-Dish Recipes* in 1904.

Two years later, Southworth published *101 Mexican Dishes*, *101 Candies*, and *101 Sauces*. In 1907 she came out with *101 Cakes*, *101 Desserts*, and *101 Ways of Serving Oysters* to round out the series.

The *Sandwiches* book started with a nice anchovy sandwich but included many surprising suggestions, including several made with flowers. The "Esthetic" involved storing unsalted butter in a jar with fresh clover blossoms before spreading the butter on thin slices of delicate white bread; the "Violet" required infusing both the butter and the bread with the floral essence; and the "Nasturtium" entailed spreading white bread with mayonnaise, laying nasturtium blossoms on the mayonnaise, and then rolling the bread around them. She also listed some sweet sandwiches, such as the Tutti-Frutti, made by chopping together crystallized cherries, peaches, and apricots and spreading on buttered wafers. Her book of *101 Sauces* ran the gamut from creole cucumber pulp seasoned with cayenne pepper to a steak sauce made from beef broth, roux, olives and sherry. Her "India" sauce for rice had a base of butter, flour, curry-powder, and stock, to which she added chopped onion, green pepper, currant jelly, and grated coconut.

Southworth's *101 Mexican Dishes* was the first cookbook in California to take Mexican cuisine as its avowed topic. She included sections on Soup, Fish, Meat, Fowl, Vegetables, Meat Dumplings, Desserts, Enchiladas, Tamales, and *Olla Podrida*. That last was not, as the name suggests, a section on stews but just a droll replacement for "Miscellaneous," covering a variety of eggy, cheesy, or bread-related recipes such as waffles, polenta, *pan relleno* (filled bread), and fried tortillas.

When compared with Pinedo's recipes, Southworth's versions of the same recipes seem less promising. At times, she abandoned clarity for cuteness, as when she named a simple rice pilaf "Mañana-land." For Spanish tripe, Southworth said to "boil the tripe until tender, and cut into narrow strips," then cook for fifteen more minutes in a sauce made from olive oil, tomato, onion, garlic, and half a chili pepper. Pinedo said to cut the tripe into strips and then into smaller, uniform pieces; then boil it for three hours with calves' feet and a cup of hominy; then add toasted chili, garlic, onion, and mint, and cook it for six more hours on a slow fire. Pinedo's insistence that the dish cook for a full nine hours augurs much better than Southworth's vague "boil until tender." Again, with beef stew, Pinedo had it cook five or six hours, where Southworth suggested only a bit more than three hours. Pinedo flavored her Spanish-style braised meat with garlic, onion, tomatoes, and chilis, where Southworth added raisins, olives, and thyme as well.

Southworth's recipe for pink beans, however, was far more detailed than Pinedo's version. The readership for Southworth's book of *101 Mexican*

Recipes had not grown up making and enjoying this dish. Southworth said to soak the beans overnight, then add onion, boil until soft, then remove the onion and add the drained beans to a skillet of fresh, sizzling-hot lard. Readers were then instructed to "mix beans and lard thoroughly until each bean seems to have a coating of the fat and begins to burst." After that, one added back the liquid in which the beans cooked and "gently crush[ed] a few of the beans to thicken the gravy." Finally, one added a chopped chili pepper and simmered until the beans were ready. In contrast, Pinedo said simply to cook the beans in boiling water, cover and let them simmer on a moderate fire, stirring from time to time without crushing them, and letting them cook until very well done. Southworth seemed to know what steps her readers would need spelled out and what ingredients they would find appealingly novel or exotic.

The Refugees' Cook Book (1906)

After the 1906 earthquake and fire, safety concerns about damaged flues meant that many San Franciscans found themselves cooking on the street outside their homes. Cooking became a public activity, and people would stand around and discuss what a neighbor was making. At that time, an enterprising group of women funded the publication of a small cookbook, *The Refugees' Cook Book: 50 Recipes for 50 Cents*, compiled by a twenty-two-year-old nurse named Hattie P. Bowman, who herself was apparently one of the refugees. She wrote in the preface that the book was intended for those who had lost all their own cookbooks in the fire:

> Every recipe has been tried and tested and in the next edition there will be many useful suggestions and more recipes, explained so understandingly that any one, no matter how inexperienced, can prepare without difficulty.[10]

The next edition would have easy recipes for inexperienced cooks; this one claimed only that the recipes had been tested—no word if they were tested in temporary outdoor kitchens. Inexperienced cooks in makeshift kitchens might be understandably nervous when they opened to the first page of recipes and saw her instructions for making handmade noodles for soup, rolled out on a floured board "over and over" and "cut very fine." Most refugees probably just waited until the Italian groceries were open and selling noodles again.

Other dishes demanded little in the way of equipment, ingredients, or expertise. With a stove, a pot, and a kettle, one could make "Refugee Stew"—if one did not mind standing nearby to supervise the stew for three hours. The

Figure 6.1.   Cooking in the street, 1906. Courtesy of the California Historical Society, San Francisco.

beef cooked very slowly, with boiling water added as necessary to keep the meat covered. In the last hour, the cook added carrots, turnips, potatoes, parsnip, parsley, and bay leaf, and thickened the stew with a little flour and a teaspoon of caramel. With clams and sweet potatoes, as well as an oven-safe dish and an oven, one could bake a Clam Creole. With oysters, fresh corn, onion, tomatoes, and some herbs, one could easily bake an Oyster Pie. If readers had no oysters but did have a pan and a stove, Bowman suggested the following:

> **Corn Oysters:** Grate the young corn into a dish—to one pint add one egg well beaten, one small teacup of flour, one cup of cream, teaspoon of salt, drop into a hot buttered pan a teaspoonful at a time and brown well.

These pancakes probably got their name because of their puffy shape; but perhaps they also served to remember happier times, enjoying oysters as part of the good life in San Francisco.

For those with somewhat more equipment and ingredients, Bowman provided an assortment of options. The book had an enticing recipe for Rabbit Curry, for cooks who had a rabbit, and also bacon, curry powder, curry paste, and a pint of good stock. The recipe for Spanish meatballs was feasible if one had a way to grind veal, and had cinnamon, raisins, canned tomatoes, and chili peppers, in addition to basic staples such as bread, milk, sugar, lard, and onions. Potato salad only required a pot for boiling the eggs and potatoes, some celery, green onions, and Eddy's mustard, and the ability to "make a mayonnaise dressing" without further instruction.

Ambitious refugees who expected visitors might consider making this dessert:

> **Company Cake:** Two teaspoons yeast powder, sifted into three cups of flour; four eggs well beaten, two cups sugar, one-half cup cold water; one-half cup melted butter; keep out one egg and add last.

It is hard to imagine readers finding that recipe helpful, whether or not they knew how to bake in an outdoor oven. Hosts with access to an ice chest might be better off preparing Bowman's simple recipe for peach ice cream: "Take white peaches, peel and mash, sweeten to taste; add one pint of cream and one pint of milk; freeze." She also called for peaches in her recipe for mince pie:

> I always put up peaches especially for this purpose, using to seven pounds of peaches three and one-half pounds of sugar and a pint of vinegar; heat and strain; pour over fruit three mornings in succession; the last morning boil juice a little longer, adding cinnamon, cloves and mace tied up in a bag.

Those instructions read almost as an elegy for the jars of spiced peaches she must have lost in the earthquake.

Her book ended with some safety tips for the refugees, advising them to filter their water by taking an empty tomato can, poking holes in the bottom, then placing a piece of cotton batting inside with a few tablespoons of pulverized charcoal on top. After that, they could pour in boiling water, wait for it to drip through the filter, and then the water would be clear and safe to drink. She noted helpfully that a tablespoon of cayenne pepper burning in a pan would help chase away flies from one's house or tent. People in San Francisco were living through a crisis, figuring out how to avoid contaminated water and pests, but also wishing to recover a sense of normality. They looked forward to times when one might have visitors and bake a cake to be sociable. In the meantime, perhaps burning cayenne pepper worked to ward off inopportune visitors as well as flies.

Panama-Pacific International Exposition Cookbooks (1915)

After the recovery from the 1906 earthquake and fire was well underway, city leaders wanted a way to demonstrate to the rest of the country and the rest of the world that San Francisco had recovered from the crisis. The city was ready for tourists; now it had to let the tourists know. The opening of the Panama Canal provided the opportunity, and San Francisco won the privilege of celebrating the new shipping link with a spectacular world's fair. The earlier success of Chicago's 1893 Columbian Exposition led San Franciscans to envision reintroducing their city to the world on a grand scale.

Planners of the Panama-Pacific International Exposition made food a highlight of the fair. They wanted people to understand the vast developments in California's agriculture, horticulture, and food industries and to experience the cosmopolitan pleasures of San Francisco's cuisines. The canning industry seized the opportunity to explain its hygienic modern procedures, conforming with Pure Food laws. Attendees from all over the world enjoyed the gustatory delights in California's own exhibits, from big black pickled olives, to sugar plums, pomegranates, and *feijoas* (a green, egg-shaped fruit with creamy flesh and a gelatinous center). One hybrid strawberry produced a fruit "which, steaming from between two broad, hot slabs of butter-soaked biscuit dough, would make the average American family forget the mortgage and all the world besides."[11] As one visitor said: "In California, the epicureans are not the favored few, but the great democracy."[12]

From hot dogs to canned tamales, from Chinese noodles to Ghirardelli's chocolate, many snacks were available all through the day. Various restaurants offered hot roast beef, pork, and lamb sandwiches; frankfurters and hamburger steak; chop suey and Cantonese-style squab; clam chowder and ham omelets; chili con carne and tasty enchiladas; and spit-roasted chickens. The brand-new Republic of China sent exhibits to this world's fair, adding to the impression that San Francisco was an entry point for many Pacific Rim goods, including foodstuffs such as sesame oil and canned bamboo shoots.

Among the souvenirs visitors could purchase were two cookbooks released in conjunction with the international exposition. Joseph Charles Lehner called himself "the American Gastronom" and collected original menus from around the world. Lehner displayed them at the Panama-Pacific

Figure 6.2. Chinese restaurant at the Panama-Pacific International Exposition, 1915. Courtesy of the San Francisco History Center, San Francisco Public Library.

International Exposition and also had a collection of them published in San Francisco, "a city noted the world over as a mecca of epicures and gourmets."[13] The menus included presidential banquets and European royal banquets as well as several from the Palace Hotel in San Francisco. Among the latter were the Christmas dinner menu in 1902 (illustrated with a cartoon of a Chinese waiter holding a Christmas roast turkey and a Chinese lantern), and the menu from the historic reopening of the Palace Hotel in December 1909 (with dishes such as *Salade de Saison*, and a savory *Marquise à la Californienne*.) Their inclusion suggested the Palace Hotel was of worldwide significance. Lehner also included the menu from a special "Electrical Dinner" hosted at the San Francisco exposition by the General Electric Company, in honor of Thomas A. Edison, Henry Ford, and their wives. The recipes Lehner included at the end of his book did not, however, reflect any local San Francisco flavor.

In contrast, many of the recipes compiled by L. L. McLaren in his Panama-Pacific cookbook had a Californian connection.[14] He listed a "Californian" artichoke puree; a "San Franciscan" crab soup; a "California Salad" made with grapefruit pulp, peeled grapes, canned pineapple, and a French dressing; a "Fruit Fudge-Californian," made with figs, raisins, nuts, and candied cherries; a recipe for cioppino, and one for Pisco Punch. He also had dishes tied in to the exposition itself, including "Eggs à L'Exposition" and an outrageous "Pan-Pacific" salad made with shredded chicken breast, cubed pineapple, grapefruit, orange, banana, artichoke hearts, asparagus, and avocado ("alligator pear"), all mixed with highly seasoned mayonnaise.

> **Eggs à L'Exposition**: Break eight eggs into a bowl and beat slightly with a fork; add half a teaspoon of salt and the same of chile powder; then stir in a cup of fresh American dairy cheese cut into a third of an inch cubes; melt a tablespoon of butter in an omelette pan, turn in the eggs and cook on a very slow fire, stirring thoroughly until the cheese is melted and the eggs are cooked.

Mrs. De Graf's Cook Book (1922)

In 1922 Belle De Graf published a cookbook based on her experiences as editor of the *Chronicle*'s Cooking Information page. A director for the California Prune & Apricot Growers Association, De Graf made a point of including numerous recipes promoting California's produce: a salad made from California pears; sandwiches made with California raisins and walnuts; and a "California Punch" made with apricots, oranges, lemons, and apple cider. She praised the local growing season several times, noting, for instance, that one could eat lettuce every day, all through the year. She advised mothers simply to serve it with a different dressing each night: "In this manner variety is introduced and the family will not tire of lettuce."[15]

Compared with other early-twentieth-century cookbooks with advertisements, *Mrs. De Graf's Cook Book* had very few: only eight, all collected at the end of the book. But half of them included a personal declaration by De Graf that this product was the best suited for the job. An advertisement for Ghirardelli chocolate cited her saying: "[I]t is hard for me to understand why any housewife will crowd her pantry with cake chocolate for baking and cooking, and cocoa for drinking, when Ghirardelli's serves every purpose." An ad for the James Graham Manufacturing Company said: "Mrs. Belle DeGraf—noted cooking authority and culinary expert—invariably uses a Wedgewood range to test and prove her famous recipes." An advertisement for the Sperry Flour Company mentioned several products along with a very enthusiastic assessment from De Graf, who had worked at the Sperry company for many years. And the last page of the book was a full-page ad detailing "Mrs. De Graf's Endorsement of Goldberg Bowen & Company":

> All my life I have purchased fine food products of Goldberg Bowen & Company. . . . My habit of going to this oldest grocery house in San Francisco has been confirmed with every experience. . . . When I have had some doubt what product or ingredient it might be best to use, I have depended on Goldberg Bowen's advice and have always found such guidance most helpful and dependable. . . . Long ago I learned to think of Goldberg, Bowen & Company whenever I have needed anything unusual . . . and never have I been disappointed in finding what I sought.

De Graf knew that her own opinion had a great deal of value, particularly for people buying her book. She picked a few products she believed in and put her name directly into their promotional material, going beyond simply mentioning their products in the volume.

Collected Recipes of the Alcatraz Women's Club (1952)

From the title, one imagines female prisoners cooking up moonshine in their cells or concocting favorite treats out of institutional ingredients. In fact, the Alcatraz Women's Club was formed by the wives of Alcatraz's guards.[16] These women often felt isolated after moving with their husbands to the island prison, and they put together a series of community cookbooks as a fundraising project for their social club. This collection reflected the cooking of the 1950s, with its heavy emphasis on mayonnaise, gelatin, and canned produce. The Chocolate Mayonnaise Cake recipe mixed chopped dates and nuts with sugar, salt, cinnamon, cocoa, and flour, before blending in a cup of Best Foods Mayonnaise and a teaspoon of vanilla. The tomato ring salad was made of gelatin, cream cheese, tomato soup, celery, bell pepper, mayonnaise, and shrimp; the pineapple and carrot salad called for orange or lime Jell-O as well as grated carrots, canned pineapple, and a jar of maraschino cherries.

The cookbook also provided an opportunity to draw attention to the diversity of dishes eaten by the guards' families: Italian antipasto, ravioli, and pizza; stuffed peppers, enchiladas, and tamale pie; southern-style ham and egg pie; English-style Banbury tarts and Yorkshire pudding; Swedish meatballs and Swedish pancakes; and Chinese dishes such as sweet and sour spareribs, chow mein, fried rice, Chinese veal, and an Asian inspired "Springtime Skillet" that combined ground beef, rice, diced vegetables, and soy sauce. The guards' wives used these cookbooks to convey their collective cosmopolitanism, despite the relative isolation they endured while stuck on "The Rock."

Jewish Cookbooks (1909, 1931, 1987)

In 1909 the San Francisco Section of the National Council of Jewish Women put out the city's first Jewish cookbook, the *Council Cook Book*. In 1931 the Emanu-El Sisterhood published their own community cookbook: *Soup to Nuts: Cook Book for Epicures*. And in 1987 Congregation Sha'ar Zahav put out the country's first gay Jewish cookbook, called *Out of Our Kitchen Closets*.[17] All three books offered a mix of Ashkenazi Jewish dishes, such as herring salad, cabbage salad, kugels, potato pancakes, and pot roast, alongside local Bay Area favorites, including dishes with a Chinese or Mexican flavor.

The *Council* book had homemade chicken tamales and enchiladas, as well as a dish combining canned tamales with macaroni and a chili sauce, and a spicy dish called "eggs à la Mexicana." *Soup to Nuts* offered "Mexican rice," "Mexican Quacamale" ("An excellent way to make one avocado serve four to six people"), and canned tamales with chicken, as well as chop suey and "Ba Bau Fan (Chinese Eight Precious Pudding)." *Out of Our Kitchen Closets* included chili, cheese, and tofu enchiladas and a kasha tostada, as well as chicken made with sesame seeds and a Chinese plum sauce. All three also included Italian dishes, from spaghetti and ravioli in 1909 to risotto, gnocchi, and pasta puttanesca in 1987. Another trait all these recipe compilers had in common was a love of New York–style dill pickles. Surprisingly, the earlier two books did not include a single recipe for chicken soup, perhaps because they thought everyone who might buy the book already had a favorite. In contrast, the 1987 book included three different recipes for chicken soup, including one made with lima beans.

The earlier two books included shellfish recipes, ignoring kosher requirements. The *Council Cook Book* offered clam chowder made with canned tomatoes, steamed mussels, stewed lobster, and half a dozen recipes each for shrimp, crab, and oysters. Twenty-two years later, *Soup to Nuts* suggested three different ways to make lobster, three ways to make shrimp, a dozen ways to make oysters, and another dozen ways for crabs. Both books also had several recipes for pork products. Bacon did not appear as an ingredient, but the tolerance for shellfish and pork shows neither book presented itself as kosher. *Out of Our Kitchen Closets* also did not call itself a kosher cookbook, but it included no shellfish and no bacon or other pork products. When it called for sausage, the book took care to specify "beef sausages." As far as the Sephardic tradition, the books published early in the twentieth century limited themselves to a few lamb dishes and some desserts made with dates. *Out of Our Kitchen Closets* included both Iraqi-style and Greek-style haroset; as well as a garbanzo bean spread, lentil soup, and a tabouli salad.

Aside from their contribution to the cuisine of San Francisco Jews, each book was also a product of its period. The *Council Cook Book*, like other early-twentieth-century cookbooks, had many advertisements educating its readers about new products and listing suppliers happy to help provide them. A thick advertising section informed readers about grocers Lazar Bros. and Goldberg, Bowen & Co., tropical fruit importers Omey & Goetting, dairy producers Solomon Bros., fishmonger E. Bonfigli (dealer in "Fish, oysters, terrapin, frogs, crabs, lobsters and shrimps"), confectioners Lechten Bros., and many others. The book's preface thanked the advertisers, noting that "from personal experience we know their true worthiness."

The *Council Cook Book* listed common contemporary recipes for sandwiches and canapés, puddings, frozen puddings and sherbets. It also instructed readers in how to make both mince meat and mock mince pie, and presented relatively new ingredients such as canned and fresh pineapple, graham flour, Grape-Nuts, and imported ginger.

> **Council Canapés**: On slices of toasted bread, put pieces of tongue or cornbeef. Spread with mushrooms and tomatoes that have been cooked together, and sprinkle with Parmesan cheese. Put in a hot oven for five or ten minutes and serve on a hot platter.

> **Gingered Pears**: Pare, core and cut small eight pounds hard pears (preferably the fresh green Bartlett variety), half as much sugar, quarter pound Canton ginger. Let these stand together over night. In morning add one pint water, four lemons, cut small. Cook slowly for three hours. Pour into small jars. Seal when cold. Keeps indefinitely.

The Emanu-El Sisterhood's *Soup to Nuts: Cook Book for Epicures* did not have any advertisements, noting proudly that the book sold well, even beyond the immediate region. It mentioned that "the recipes were authenticated by a Special Committee which worked indefatigably to procure unusual or favorite dishes." The word "authenticated" suggests a tension between, on the one hand, the everyday foods Jewish families made at home (or had made by their cooks) and, on the other, the idea of an authentic ethnic Jewish cuisine, which young families might want to rediscover through the use of these recipes.

The many different dishes prepared in rings, timbales, and ramekins were clearly contemporary favorites. The book offered a ham timbale, a fish timbale, and a second "Surprise" fish timbale; it also listed a dish of fish "molded with cucumbers," a shrimp mousse, and a shrimp soufflé. *Soup to Nuts* listed the following "rings": artichoke, asparagus, carrot, green pea, mushroom, corn, spinach, noodles, rice, tamale, and tomato. Other recipes included tomato jelly salad, tomato mousse, corn soufflé, rye bread soufflé, and something called "cheese sponge," prepared in ramekins.

Tamale Ring: chop together 1 bell pepper, 1 clove garlic, 1 pint pitted olives. Mix with 1 small can tomatoes, 1 cup yellow corn meal, 1 cup milk, salt, 1 heaping tablespoon Grandma's pepper, 2 beaten yolks. Add stiffly beaten whites. Pour into buttered ring mold in a pan of hot water and bake 1 hour in medium oven. Fill with creamed chicken, shrimps, etc.

Rye Bread Soufflé: 3 eggs, 1 cup sugar, 1 tsp each cinnamon [and] cloves, 1 cup grated rye bread crumbs. Beat yolks and sugar. Add crumbs and spices, then stiffly beaten whites. Butter baking dish and bake in oven 20 minutes or more. Serve with cream.

Ba Bau Fan (Chinese Eight Precious Pudding): 1 lb glutinous rice, 3 oz dates, 1 oz dried cherries, 1 oz candied watermelon, ½ lb sugar, 2 oz lotus seeds, 1 oz green plums, 1 oz. barley (cooked). First cook the rice; when done mix in the sugar. Oil Chinese bowl with lard, and then put into it in layers and in rows, all of the ingredients. Fill the bowl with the rice and steam for two hours. When ready to serve invert pudding into a large plate. Serve hot.

Potato Pancakes: Grate four potatoes, suspend in napkin to remove all water. Then add almost one pint of boiling milk to dried potatoes, 1 tablespoon flour, 2 eggs, 2 grated apples, salt. Fry like pancakes until crisp. Serve very hot.

More than fifty years later, *Out of Our Kitchen Closets* reflected a very different sensibility. A notation on the copyright page specified that three dollars from every copy sold would go to the Food Bank of the San Francisco AIDS Foundation, "providing food and domestic supplies free of charge to AIDS patients." And the cookbook suggested menus for readers' coming-out cocktail parties, pre-Pride Parade buffets, and a lawn lunch for one's commitment ceremony, as well as menus for a bar or bat mitzvah reception, a Rosh Hashanah Eve dinner, and other typical Jewish events.

Written in the 1980s, this gay Jewish cookbook also included many vegetarian dishes and quiches, as well as brownies, Kahlua pecan pie, kosher chocolate mousse made with Kahlua, sixteen different kugel recipes, and a section on "Noshes." The book referred back lovingly to familiar recipes from childhood, including "Mama's Kugel," "Nana's Egg Twist," "Mother's Kreplach," and "Nana Hattie Schiltz' Tiny Meat Blintzes." The headnotes mentioned family lore such as "My mother is famous for this brisket." Whether for gefilte fish or vegetarian tsimmes, the recipes gave very detailed instructions, exhibiting the increased professionalism of recent cookbooks. Many gay readers had not learned to cook alongside their mothers and grandmothers and probably appreciated the book's clear explanations. A recipe titled "Cheese Soufflé That Ain't" was pretty basic. The instructions said to layer cubes of thick sourdough bread with grated Swiss cheese in a casserole, and then pour a mixture of eggs, mustard, fine herbs, garlic powder, and Worcestershire sauce over the top. The headnote was reassuring: "Afraid to try a soufflé? Don't be. This one is guaranteed not to collapse." Gay people were making a life for themselves in San Francisco, complete with both favorite childhood treats and dishes to share with a new community.

Russian Cookbooks (1973, 1974)

The early 1970s saw the publication of two cookbooks, one a community cookbook that originated within the Women's Circle of San Francisco's First Russian Christian Molokan Church (1973), and the other published the next year by Margaret Koehler, based on her experiences visiting Russians in the Inner Richmond district.[18] Both books agree that Russian immigrants changed their diet dramatically after arriving in San Francisco. They adapted quickly to using a vast array of prepared and frozen foods rather than the more limited ingredients available back in southern Russia.

The Women's Circle book emphasized cooking from scratch as much as possible, whereas Koehler's cookbook insisted that "short cuts are perfectly permissible. . . . Russian housewives in San Francisco regularly use convenience foods in their recipes."[19] Thus, for *piroshki*, the Women's Circle explained how to make the dough with flour, eggs, sugar, salt, butter, oil, yeast, and hot milk, whereas Koehler recommended starting with a hot roll mix; again, for blini, the Women's Circle started with basic ingredients, whereas Koehler started with packaged pancake mix. For Chicken Kiev, the Women's Circle version started by making an herbed butter with parsley, tarragon, garlic, salt, and pepper; Koehler inserted plain chilled butter inside rolls of flattened chicken before frying the rolls and serving on toast. The Women's Circle also suggested one should make one's own *pelmeny*, filling the dumplings with a mix of ground beef, lamb, onion, and garlic. In contrast, Koehler said to buy them frozen from Russian grocery stores or else have them delivered to one's doorstep:

> One church group has made quite a successful fund-raising enterprise by making pelmeny, freezing them, and then supplying them to homes. While in San

Francisco recently, I woke up one morning to hear much Russian conversation going on out in the hallway. When I asked my hostess who the early visitor had been, she said "The pelmeny man. He comes once a month."[20]

Both books assured readers that they could make beef Stroganoff from inferior cuts of steak. Koehler noted that if one cut the steak into long, thin strips across the grain, the pieces would end up more tender, almost as good as a boneless sirloin. She suggested that the "real Russian way" to serve this dish was with fried potatoes, noting that "frozen shoestring potatoes are perfect."

Both cookbooks stressed the importance of *zakuski* (appetizers) when Russians get together socially. Koehler suggested providing at least one kind of *zakuski* for every two guests. Both books recommended caviar, herring, *pashtet* (liverwurst), a vinaigrette salad, cucumbers with sour cream, pickled mushrooms, cold cuts, and an assortment of breads, along with well-chilled vodka. Their stuffed egg recipes varied: the Women's Circle recommended mixing the hard-boiled yolks with sour cream, butter, and dill; Koehler proposed filling the egg white halves with caviar instead, for a more extravagant effect. She also suggested that instead of pickling fresh button mushrooms in vinegar and tomato paste, one should buy canned button mushrooms, pack them in a jar with bottled Italian salad dressing, and refrigerate them for a few days to achieve a very similar result. Likewise, one could make Russian potato salad by buying a prepared potato salad and adding beets, pickle, onion, and chopped egg: "This has a very pretty rosy color and a delicious flavor." Even a simple store-bought jar of pickled beets could serve as a "perfectly acceptable zakuski."[21]

The cookbooks included a variety of borscht recipes, with meat and without, with beets and without. Chicken borscht was made with chicken backs, cabbage, potatoes, and cans of tomato sauce, whole tomatoes, and stewed tomatoes. The books told how to make green *stchee*, cooking beef brisket for an hour or so, then adding chopped spinach for the last few minutes, and topping with sour cream and egg. Another soup described in loving detail was *rassolnik*, which took its distinctive flavor from veal or beef kidney. They also offered a wide variety of traditional desserts, from *kissel* and *khvorost* (puddings and fritters) to *drachena* and *vatrushka* (sweet omelets and sweet dough rings), and others, including Russian tea cakes. Tea was an important part of Russian culture. Both books explained the key role of the Russian samovar; they also noted that San Francisco's Russians tended to use a kettle or an electric samovar to heat the water for their tea, even if they displayed a beautiful old-fashioned samovar on a sideboard.

The Women's Circle went beyond ordinary recipes for the household's meals and entertaining. These women explained how to can tuna and

salmon; how to make the fermented drink *kvass* from black rye bread; and
how to cook enough food for a large church dinner. In order to make borscht
for three hundred people, for instance, one started the day before by chopping four bunches of beets, fifteen heads of cabbage, and assorted other fresh
vegetables; one also set fifteen pounds of soup bones to soak overnight. The
morning of the event, the work began at 4:30 a.m., setting 250 pounds of
meat to boil in four huge kettles. Then one began peeling thirty pounds of
potatoes and eight pounds of onions. By 7:30 a.m., one took the meat out of
the kettles and began roasting it, and in just a few hours one could add the
vegetables to the beef broth and bring the borscht to a boil in the enormous
kettles.

Chinese Cookbooks (1910, 1913, 1963, 1974, 2007)

In the 1910s two cookbooks were produced on Grant Street in San Francisco's Chinatown. Both aimed to help literate Chinese men cook for their
English-speaking employers (in households or in restaurants), and both were
referred to in English simply as "Chinese and English Cook Book." Neither
book included any Chinese dishes, unless one counts concise instructions for
brewing tea. Instead, the books instructed Chinese cooks on how to make
British colonial dishes such as mulligatawny soup and Indian curry; southern
American dishes such as hominy, stewed okra, and fried tomatoes; as well as
a diverse range of dishes such as chili sauce, corned beef, matzo pancakes,
Irish stew, Hungarian goulash, and chicken Marengo. Just a few years later,
across the continent in New York City, Shiu Wong Chan broke away with
an entirely new approach. His *Chinese Cook Book* contained "more than one
hundred recipes for everyday food prepared in the wholesome Chinese way."
His book encouraged New York housewives to cook and eat Chinese food
themselves; a bold new project compared with the San Francisco books of
the same decade, which merely helped Chinese people serve Western food
to their employers.

   In San Francisco, the first cookbook introducing non-Chinese people
to Chinese cooking was Johnny Kan's *Eight Immortal Flavors* (1963). After
decades running restaurants in Chinatown, and ten years at his landmark
Kan's Restaurant on Grant Avenue, Kan realized that Americans were
hungry for Chinese food but had many misconceptions about the cuisine.
Americans tended to order main dishes without any rice; they also had
begun buying frozen and canned Chinese foods but did not know how to
incorporate them into a Chinese-style meal. Kan also wanted to provide
a resource for the "American-born Chinese housewife who has forgot-

ten how to cook the dishes her mother once prepared."[22] And he wanted to educate both American and Chinese-American housewives in the importance of taking cooking seriously, giving proper attention to even the smallest steps, and appreciating good, simple food. He explained that children in China learned first to wash rice and then cook it properly, and they learned to value the few flavorings they had, such as peanut oil, salt, and soy sauce.

Kan and his coauthor, San Francisco-based journalist Charles Leong, were both Cantonese in origin, so it was natural for them to present their readers with an education in Cantonese cuisine. Within southern Chinese cooking, Kan explained techniques such as "toss-cooking" (stir-frying); "smother cooking" (braising); and the importance of steamers and woks. He had a section on cleavers and the different cuts necessary to mince, dice, shred, or slash various ingredients before cooking. He included drawings of many common products, from star anise to lotus seeds and soy bean skin, and vegetables from winter melon to long beans and Chinese parsley. And, of course, he explained the eight immortal flavors of the title:

> Memorize and recite with verve and rhythm—sing-song like—thus, in a swing of two words in each cadence:
> Hom—Tom; Teem—Seen; Foo—Lot; Heong—Gum

Or, translated: salty—neutral; sweet—sour; bitter—spicy; fragrant—tart. Kan wanted to explain the basics of the cuisine and offered a multitude of accessible dishes, devoting only a small part of the book to unusual or difficult preparations.

Each section of the book began with a discussion of the role of that food in Chinese culture, as well as possible adaptations when living in the United States. Most of the recipes ended with a short note in Kan's voice, explaining special aspects of the dish, or particularly delicious variations. For Chinese okra soup, he noted its "refreshing garden greenness and texture—unexpected in a soup"; for "Mock Lamb" he explained that the "fluffy rice sticks" around the stir-fried beef create the effect of a lamb's wool; for soy sauce chicken he suggested that children loved the chicken-flavored sauce mixed into their steamed white rice; for spareribs with black bean sauce he declared the dish ideal for taking on overnight camping trips. Kan recommended his seaweed soup by saying "you'll be surprised by its unique blend of fresh ocean flavor with a natural 'sweetness.'" In explaining "Crab a la Kan," he added with warmth, "simply 'let yourself go' and use your fingers!"—good advice for those who wanted to enjoy that dish to the fullest.

In addition to a cookbook's standard sections on soups, meats, desserts, and so forth, Kan also included an unusual category: "Chinese Tidbits," which included all the little snacks people might enjoy between meals. He provided recipes for many of these, from shrimp balls and barbecued pork to crunchy chicken livers, observing with a fellow cook's empathy or even trepidation that "the perfect Chinese host or hostess can be kept busy in the kitchen morning, noon and night with a tremendous range of dishes designed for around the clock enjoyment." Kan suggested that those who lived in San Francisco should only make a few of these dishes, such as *jook* (thick rice soup) and *lo mein*; all the dim sum specialties should instead be bought from the many wonderful Chinatown establishments. Indeed, Kan's book was in part an introduction to the joys of San Francisco's Chinatown, explaining the diverse activities going on there at different times of day.

In 1974 Cecilia Chiang worked with Allan Carr to publish a selective version of her life story, *The Mandarin Way*, with recipes tailored for American home cooks. They emphasized the importance of seasonal cooking and of using only the highest quality ingredients. In a delightful chapter on cocktail parties, Chiang explained that, in China, mandarins did not hold such events—but "the versatility of the Chinese repertoire is perfectly capable of supplying tidbits in harmony with a Western cocktail party."[23] She introduced her readers to mu shu pork, suggesting that they could serve the wrapped treats to cocktail-hour guests who stay late, after the spring rolls and shrimp balls had been passed and eaten. Another chapter was titled "Of the Unexpected Guest and the Resourceful Hostess," with helpful recommendations for preprepared delicacies such as century eggs and Chinese sausages. Chiang also suggested serving unexpected guests red-cooked pork shoulder, which would take almost two hours to cook. The Western hostess was "under no compulsion to serve six or seven dishes," so her energy could go toward making one exceptional dish.

Thirty years later Chiang and food writer Lisa Weiss began drafting a beautiful book that would reintroduce a new generation to Chiang's experiences, telling many stories that political concerns had kept out of the first memoir and sharing some of the most famous recipes from her restaurant. *The Seventh Daughter: My Culinary Journey from Beijing to San Francisco* came out the year after Chiang closed The Mandarin. The book included instructions for how to make the Mandarin's pot stickers, spicy twice-fried green beans, spicy eggplant, eggplant in garlic sauce, prawns "à la Szechwan," minced squab in lettuce cups, Mandarin Crispy Chicken Salad, and a simplified version of the restaurant's Chongqing spicy dry-shredded beef.

Along with her reminiscences from childhood, young adulthood, and her years at The Mandarin, she sprinkled in her own favorite recipes, many with particular significance: her mother's red-cooked pork, made with Shaoxing wine, ginger, and dark soy sauce; the five-spice slow-cooked beef her father enjoyed as an appetizer before dinner; and the green-onion pancakes she would eat after school. She also included recipes that provided comfort during harder times: the hot-and-sour cabbage the family subsisted on during the Japanese occupation, the congee her mother made before sending Chiang and her sister on an arduous journey to the unoccupied part of China, and the *dan dan mian* noodle dish she wolfed down as street food after she and her sister finally arrived in Chongqing. Chiang's personal recipe for garlic noodles used just fresh noodles, garlic, a little oyster sauce, and even less soy sauce. The dish is "so simple it's almost embarrassing," in Chiang's words— too simple to go on The Mandarin's menu—and yet quickly became one of the most popular dishes for people to order off-menu.[24]

Restaurant Cookbooks

The trend toward restaurant cookbooks began long before Chiang published her recipes from The Mandarin. In 1910 Victor Hirtzler, head chef of the St. Francis Hotel, began promoting the hotel by putting out books listing elaborate menus for breakfast, lunch, and dinner, all three meals listed for every single day of the year. Readers unprepared to follow his plans for each day could rely on the extensive multipart index. The 1919 edition, *The Hotel St. Francis Cookbook*, became very well known. The sheer exuberance of the diary format, squeezing so many recipes in so tight a space (only a few overflowed their day's page), gave the cookbook a certain charm, complementing the prestige of Hirtzler's culinary skills. Like Belle De Graf a couple of years later, Hirtzler was happy to promote California's produce, proposing on October 23rd, for instance, sliced oranges, grapefruit, and strawberries; "Prunes Victor"; avocado with French dressing; and chicken stuffed with raisins. He noted that "California raisins may be used in many dishes, such as soup, fish, entrees, roasts, bread, puddings, ice cream, etc."[25]

Then, in 1963 Doris Muscatine published her famous book, *A Cook's Tour of San Francisco: The Best Restaurants and their Recipes*, and San Francisco has never been the same. With her inimitable voice, she took readers on a tour of her favorite restaurants, all across the city, visiting all genres of eateries. After starting with a historical overview of San Francisco cuisine, she offered some suggestions for cheap bites, from sourdough bread at the French Bakery, to "excellent burritos with sour cream" at Tia Margarita on Nineteenth Avenue, to red beans and rice (and jazz) at Earthquake McGoon's on Clay Street.

A few pages later, she was enjoying the specialties at Jack's, including calf's head vinaigrette on Thursdays. Next, she and her readers were luxuriating over at the Fleur de Lys; owners Cherie and Robert Charles generously let Muscatine print their recipes in her book: Provençale fish soup (calling for six different kinds of saltwater fish as well as shellfish, onion, thyme, laurel, fennel, and a cup of the Charles's aïoli); frogs' legs "Giselle"; and coeur de filet Périgourdine, which required two pounds of goose liver and a pound of imported truffles in addition to the "choice filet of Eastern steer." Muscatine then shifted to the restored Ritz Old Poodle Dog restaurant and suggested readers might prepare the restaurant's chicken à la Ritz, warning them to serve the dish immediately after adding two egg yolks, with no further cooking. She steered her audience to elegant Japanese restaurants as well. At Yamato's on California Street, the oldest Japanese restaurant in San Francisco, she acquired a very short recipe for lobster sashimi and a very long recipe for beef sukiyaki, in her view probably the "tastiest of the Japanese dishes."

She loved more than just the high-end restaurants, however, and in Chinatown she advised her readers to go to the Sam Wo *jook* house as well as to Kan's elegant restaurant. At Sam Wo, they could get the best *jook* and the best Chinese crullers, while being teased by longtime waiter and charming San Francisco character Edsel Ford Fung. Diners could also enjoy marinated raw fish salad, for which she provided the recipe accompanied by a note to any unadventurous readers: "If you are not up to raw fish, cold, boiled fish can be used, though Edsel would point out with some measure of disappointment that this was neither authentic nor exotic." Edsel Ford Fung died in 1984, and Sam Wo itself closed in 2012.

Muscatine also recommended that her readers pass Trader Vic's *rumaki* hors d'oeuvres at their cocktail parties: a strip of bacon wrapped around a bite-sized piece of cooked chicken liver and a bit of crunchy water chestnut, deep-fried to the point of crispiness. With evocative imagery she brought readers to Sam's Anchor Café in Tiburon, north of the city, and gave them Sam's recipe for crab cioppino. She even persuaded the Yugoslavs running Maye's Oyster House on Polk Street to give her their recipes for Hangtown Fry and Oyster Loaf; the latter recipe called for hollowing out a loaf of French bread, filling it with fried oysters, and topping it all with a pile of French fries.

In 1981, almost twenty years after Muscatine's book first came out, Brian St. Pierre and Mary Etta Moose put together a more specialized volume with a similar approach. Like Muscatine, the authors adopted a playful, personable voice as they explained the history, gave expansive restaurant reviews, and included some favorite recipes for readers to try, but this time all from one neighborhood.[26] *The Flavor of North Beach* starts with a vast glossary, fourteen pages, from Al Dente to Zuppa Inglese, by way of terms such as the following:

Carciofi (Artichokes): The Italians prize the smallest artichokes during their brief season, July through August. The baby artichoke appears in the local markets from ¾ to 2 inches in diameter. After the prickly tip has been whacked off, the rest is so tender it can be cooked and eaten whole, embryonic choke and all.

Cioppino: Time was you could expect your local to be serving cioppino (crab stew) to accommodate its Friday night drinkers, whether the saloon served food or not. Our Dungeness crab was plentiful, and the young fishermen and their hot girls presented a Tom Jones scene, standing around the dance floor sucking crab out of their shells over steaming bowls.

Sacripantina: A North Beach dessert made exclusively by Stella Pastry. The word is said to translate, loosely, as "Oh, how marvelous!"

After exploring their way through some thirty restaurants and a handful of coffeehouses, butchers' shops, bakeries, and Italian delis, the authors presented a section of recipes and anecdotes.

Some of their recipes came with warnings, as in the extended note accompanying Chef Ron Barber's Vitella Nordbicciana:

Like many good restaurant dishes, this one is prepared—occasionally before your eyes—in a deceptively quick and easy manner. Part of the deception is that half-cup of demi-glacé, the sine qua non of the good restaurant, folded in at the last minute. (You'll need half a day in the kitchen to make an unsalted veal stock to reduce into a glacé, unless you know a chef who moonlights selling demi-glacé.) . . . (I'd advise having a standby steak at the ready if your first try is veal for a company meal.)

In other words, do not try this at home, or at least not when you have company coming. The detailed instructions for making crab cioppino are almost sufficient to persuade the reader to give it a whack, with a sharp cleaver:

Chef Manuel Sausedo's Crab Cioppino and Polenta: The North Beach chef's secret to a proper cioppino is to dissect the crab alive, and to so cut each leg that the section of body from which it protrudes is left attached to the leg. These two steps keep the crab juicy during sautéing, and give the stew the intense crab flavor that makes this so different from other fish stews. . . . Lay the crabs on their backs on a table, and, with one whack of a sharp cleaver, bisect them from head to tail. This dispatches them more quickly than throwing them into boiling water, which will also detract from their flavor. Remove the back shell and the gray fibrous matter (do this over a bowl to catch all the yellow-orange stuff and white crab fat, with which you will flavor the stew.)

Most readers would instead look for an excuse to relax at Caffe Malvina or Tosca Cafe nearby, and let experts make a proper crab cioppino.

Fine dining experiences do often motivate people to buy cookbooks from favorite restaurants, although readers may not actually prepare those recipes in their home kitchens. Quite a number of well-loved cookbooks have come out of other Bay Area restaurants over the years, such as Narsai's, Chez Panisse, Zuni Cafe, the Tadich Grill, Henry Chung's Hunan Restaurant, Omar Khayyam's, and Boulevard. Many hip guides to San Francisco restaurants have also been published, including Jack Lord and Jenn Shaw's book *Where to Sin in San Francisco* (1939); Arthur Fleming's *My Secret San Francisco: How to Eat, Drink and Swing in San Francisco on Almost No Money* (1966); and "Jack Shelton's Private Guide to Restaurants," a newsletter Shelton put out from 1967 to 1978 (his successor, Robert Finigan, kept it going until 1985).

Since the 1980s, the restaurant scene in San Francisco has become even more competitive, with chefs struggling to obtain and then maintain their reputations. Beyond the long-standing California Cuisine and locavore movements, newer urban trends from molecular gastronomy to pop-up restaurants and food trucks have made it even harder for an established restaurant to stay relevant. Recent guides to help navigate the city's changing restaurant scene include Carolyn Miller and Sharon Smith's *Savoring San Francisco: Recipes from the City's Neighborhood Restaurants* (2000, 2005) and *The Chowhound's Guide to the San Francisco Bay Area* (2005). Old or new, tried and true, these restaurant guides appealed to readers who would just as soon go out to another great San Francisco meal as cook one at home.

CHAPTER SEVEN

~

Signature Dishes

Over the decades, some dishes have become associated with San Francisco's history and creativity. In many cases, that connection was established or strengthened by the city's inimitable icon, Herb Caen, who began writing a column for the *Chronicle* in 1936 and submitted his last column in 1997.

Cappuccino

The Tosca Cafe opened in the early 1930s and may have been the first establishment to serve cappuccino in San Francisco—but the bar's "special" cappuccino was more a celebration of the end of Prohibition than a tasty coffee beverage. *Chronicle* columnist Robert O'Brien enjoyed spending time there, listening to operatic arias on the jukebox: "you sit at the unpretentious bar and sip a *cappuccino*, a fabulous concoction straight from the Apennines that is chocolate, heated by steam forced through coffee, and laced with brandy."[1] Coffee made only a minor contribution to the famous (Caen-approved) Tosca cappuccino.

Real cappuccinos did not take hold until the late 1940s. Thomas Cara came back to San Francisco and opened a kitchenware shop. Having spent World War II stationed in Italy, he recognized that returning servicemen would want to recapture the tastes they had enjoyed in Europe. He began importing Italian espresso machines, especially the top-of-the-line Pavoni brand, and sold them to restaurants and bars in San Francisco. Customers who had previously enjoyed a French-style café noir began expecting cappuccinos, especially in North Beach. Former mayor Joseph Alioto called it the "Cappuccino Circuit"—the familiar pathways that led to long conversations over coffee at Caffe Malvina, Caffé Greco, Caffé Trieste, Caffé Puccini, Caffé Roma, the North End Caffé, and Mario's Bohemian Cigar Store. A new taste for Italian-style coffee may have contributed to San Francisco's 1963 revolt against ordinary restaurant coffee, as exemplified by the famously melodramatic *Chronicle* headline: "A Great City's People Forced to Drink Swill."[2] Before long, Alfred Peet had opened his specialty coffee shop in Berkeley, teaching locals (including Alice Waters) the importance of sourcing, roasting, and enjoying fine coffees from around the world.

Celery Victor

Author Clarence Edwords once approached Victor Hirtzler, head chef at the St. Francis as well as a promotional genius in a red fez, and asked for his best recipe. The Alsatian chef first described how to make sole in a style derived from Escoffier, but then offered up a more personal recipe called "Celery Victor." It is not a particularly flashy recipe. There are no exotic ingredients to dress up the celery. It is, at first, somewhat surprising that the flamboyant Hirtzler named the dish after himself and thought to mention it to Edwords.

San Francisco's celery customs may have been a sore spot for this classically trained chef. In the early twentieth century, San Franciscans expected celery on the table among the hors d'oeuvres, and celery was always on the table at Thanksgiving. Aside from using it for stuffing or cream of celery soup, diners expected to eat celery raw. French people, on the other hand, had much less enthusiasm for raw produce. By cooking celery in a rich, meaty stock, Hirtzler tamed this crunchy vegetable and made it acceptable to a more sophisticated palate, which meant, in his view, a French-trained palate. Signaling the success of Hirtzler's approach, Celery Victor soon became fashionable at Thanksgiving and Christmas banquets across San Francisco.

> **Celery Victor:** Take six stalks of celery well washed. Make a stock of one soup hen or chicken bones, and five pounds of veal bones in the usual manner, with carrots, onions, parsley, bay leaves, salt and pepper. Place the celery in a vessel and strain the broth over it. Boil until soft and let cool off in its own broth.
>
> When cold press the broth out of the celery with the hand, gently, and place on a plate. Season with salt, fresh ground black pepper, chervil, and one-quarter white wine vinegar with tarragon to three-quarters of best olive oil.[3]

Chicken (or Turkey) Tetrazzini

This dish was named for Luisa Tetrazzini, an opera singer well loved in San Francisco. The city by the bay was where she first gained fame in 1905, before becoming an international sensation. The Palace Hotel often hosted the opera star, and Palace chef Ernest Arbogast has been called the creator of chicken Tetrazzini. He may have invented it the night of March 6, 1905, when Tetrazzini sang in the hotel's Palm Court to much acclaim. Reporters mentioned that Palace manager John Kirkpatrick gave the singer a pearl pendant at the banquet afterward; unfortunately, they did not mention a new culinary creation.

The October 1908 issue of *Good Housekeeping* has the earliest printed reference to the dish and locates it in New York:

> **Turkey Tetrazzini:** At the restaurant on Forty-second street they serve a good and easy entrée or main course. It is named after the famous singer. Small, thin slices of cooked turkey in a cream sauce to which some cooked spaghetti was added and a little grated cheese, also some very thin slices of mushrooms cut crossways. This was served in the dish in which it was cooked and some bread crumbs were browned over the top.[4]

The "restaurant on Forty-second" may refer to the Knickerbocker Hotel, then located at the corner of 42nd Street and Broadway in New York City— and often home to Tetrazzini, Enrico Caruso, and other opera legends.

In 1910 Tetrazzini's New York employer, Oscar Hammerstein (the grandfather of the famous lyricist), forbade her from singing in San Francisco. She defied his contractual claims, declaring that no one could stop her from singing in the streets, "for [she] knew the streets of San Francisco were free."[5] There she was, outside on Christmas Eve, singing "The Last Rose of Summer" in front of Lotta's Fountain by the Palace Hotel. Two hundred thousand people stood and listened, and then Tetrazzini and the crowd together sang "Auld Lang Syne." The opera star hoped that by daring to sing outside in late December, she could showcase San Francisco's wonderful weather and support its pitch to host the Panama-Pacific International Exposition.[6]

Some observers noted that Tetrazzini had put on weight since her 1905 appearances and now resembled "a prosperous citizen of some town extremely fond of good living."[7] Later in life, she got even fatter, but insisted she ate well and was in the best of health:

> I thrive on butter and eggs. As for spaghetti, I often eat it twice daily. Sometimes I drink a pint of rich cream at one meal. I revel in olive oil, not only because I like the taste, but because it is good for the muscles of the throat . . .

Figure 7.1. Lotta's Fountain on Market Street. Courtesy of the San Francisco History Center, San Francisco Public Library.

Think of trying to go through "Lucia" or "Rigoletto" on a luncheon of lettuce and a glass of lemonade—it could never be done.[8]

In November 1912 the *San Francisco Call* echoed the 1908 *Good Housekeeping* article, mentioning that "a restaurant in New York serves a delicious entree named after 'Our Luisa.'"[9] Just two years after Tetrazzini's performance in the streets of San Francisco, a local newspaper matter-of-factly referenced the East Coast origins of turkey Tetrazzini. Whether or not Arbogast created a special Tetrazzini entree, later San Francisco residents happily adopted chicken Tetrazzini with the same affection they offered the great singer.

Chop Suey and Fortune Cookies

Chop suey and fortune cookies are both associated with San Francisco, though they originally emerged elsewhere. Chop suey (*shap sui* in Cantonese) was initially a stir-fry of various vegetables and organ meats. The dish was brought to America by Chinese from the Sze Yap area, just to the south of Guangzhou. These Cantonese arrived on the West Coast, but by the 1880s many had continued on to New York City. In 1884 Wong Ching Foo wrote in the *Brooklyn Eagle* about "chop soly," and by 1886 the New York journalist Allan Forman eagerly informed his readers that chop suey was both delicious and cheap.[10]

Non-Chinese San Franciscans did not start appreciating chop suey until around 1900. The city's Chinese restaurateurs believed, however, that they had come up with chop suey on their own. In 1904 a San Francisco cook, Lem Sen, showed up in New York claiming to have invented chop suey and demanding royalties from all East Coast restaurants advertising the dish. He failed in that effort, but succeeded in highlighting San Francisco's attachment to chop suey. Stories spread that chop suey had been a Chinese cook's revenge, frying up old scraps to serve to racist forty-niners, but that only made San Francisco Bohemians love chop suey all the more.[11]

Chop Suey: Barely cover a small chicken with water and boil until tender; then shred the meat, return the bones to the soup, boil down to one cup; then strain. Cut a half pound of lean pork into thin inch strips and fry; when brown, add the chicken, a half cup of celery, cut thin, a small piece of chopped onion, six dried Chinese mushrooms (which have soaked in water), six water chestnuts, cut thin, a quarter of a pound of wheat or bean sprouts, half a chopped green pepper, and a small piece of ginger-root, crushed. Pour over all the chicken broth and season well with a very little salt and a tablespoon each of *gu yow* [mushroom sauce] and soy—sauces to be found in any Chinese provision house. Simmer for a few minutes and serve with rice.[12]

Fortune cookies were originally Japanese tea wafers (*senbei*). From 1895 to 1925, Makoto Hagiwara maintained Golden Gate Park's Japanese Tea Garden, created for the 1894 Midwinter Fair. His family served visitors "tea in small cups, unsweetened, along with wafery cakes, the edges of which are mysteriously curled and brittle."[13] By 1901 one could buy *senbei* in Y. Wakashiro's bakery on the southern edge of Chinatown:

[The baker uses] iron molds resembling tongs, and . . . a tilted bucket of thin dough of a light yellow color. Each tong contains dough for one crisp wafery "senbi" which takes at least five minutes to bake. As each cake is baked the operator unclamps the tongs, removes the soft cake, which stiffens when cold until it is brittle as a chip, and then places it on an improvised table, putting a light weight on it in order to hold it flat. . . .

The style turned out depends much on the way in which they are handled after leaving the molds. Some are rolled over a bamboo reed and shape themselves like a piece of paper which has been revolved around a pencil. . . . The "senbi" cakes in their various shapes are exceedingly fine tea biscuits. . . . They are thin and crisp, although there is a certain toughness about them. The dough used is made of common wheat flour, eggs, sugar and water.[14]

Senbei were sometimes savory treats, but Wakashiro made them as crispy cookies. In Japan, shops near shrines offered special *tsujiura senbei*—rice crackers bent around paper fortunes—but San Francisco's Japanese tea cakes did not at first include fortunes.

At the 1915 Panama-Pacific International Exposition, the Sperry Flour Company built a stunning exhibit with international breads made daily in front of the public, including tortillas, piroshkies, matzo, Chinese almond cakes, and also Japanese *senbei*. Japanese tourists came to the exposition and may have suggested adding little fortunes to the cookies. By that point, the Benkyodo bakery on Geary made *senbei* both for the Japanese Tea Garden and for the public—and also adapted *senbei* techniques to making ice cream cones (then called "cornucopias"). The very first American *tsujiura senbei* were scroll-shaped like the ones rolled on a reed at Wakashiro's bakery, with a slip of paper tucked inside: "sort of a Confucius tamale."[15]

In the 1920s Japanese companies in San Francisco and Los Angeles began marketing "fortune tea cakes" as party favors. Chop suey restaurants adopted them as a trendy and sweet giveaway, while Japanese and Chinese bakeries worked to make the little cookies sturdier. Then came World War II. The Japanese spent years in internment camps, while Chinese baking companies established relationships with Chinese restaurants across America. As soldiers came through San Francisco, they enjoyed chop suey and learned to associate Chinese food with fortune cookies. Back in their hometowns, they then encouraged Chinese restaurateurs to adopt the new "Chinese" fortune cookie.

Cioppino

Italians from Genoa arrived in California with extensive fishing experience and a taste for a thick fish soup they called *ciuppin*. The earliest San Francisco references to this dish emerged in stories about the city's picturesque fishing

Bernstein's
FISH GROTTO

SERVING SEA FOOD EXCLUSIVELY
FOR THE PAST FORTY YEARS

FISH CAUGHT AT 5 A.M. SERVED HERE THE SAME DAY!

 A PERFECT COMPANION OF SEA FOOD
A BOTTLE OF CHABLIS OR RHINE WINE

SOUPS	BOWL	CUP
Our Famous Clam Chowder	.25	.15
Cream of Abalone	.40	.20
Mock Turtle	.25	.15
Lobster, Bisque Maurice	.40	.20
Crab Gumbo, Louisiana	.35	.20
Clam Broth	.30	.20

SALADS
De Luxe Marine Salad Bowl 1.60
Stuffed Tomato with Crab or Shrimp 1.30
Crab Meat or California Shrimps, Louie 1.25
Crab Legs on Bed of Lettuce 1.60
Crab or Shrimp Salad 1.20
Jumbo Prawns on Crisp Lettuce 1.40
Tuna Salad on Crisp Lettuce 1.00
Mixed Green Salad 90
Cottage Cheese Salad .65, with Fruit 90
Crisp Lettuce and Garden Ripe Tomato 60
Heart of Crisp Lettuce and Anchovy 90
Half Avocado .65, Stuffed with Crab Meat 1.35
California Lobster Salad 1.60, Marine Combination 1.50
Neptune Sea Food Salad 1.55 .... Sliced Tomatoes 50

FROM THE
Chef's Galley

Barbecued Prawns, Penguin Sauce 1.45
The Captain's Special Platter: A Combination of Fried Oysters, Crab Legs,
     Abalone, Scallops, Prawns and Filet of Sole. Tartar Sauce 1.85
Lobster Thermidor 2.00
Lobster Princesse: Baked Lobster Meat, Heart of Artichokes, Fresh Mushrooms,
     Green Peppers, Sherry Wine adds a Piquant Touch 2.00
Pieces of Eight: Timbale of Lobster, Shrimps, Crab Legs, Sliced Avocado and
     Tomato, Buccaneer Dressing   So Different! 1.80
Crab Meat Newburg or a la King en Casserole 1.55
Eastern Oysters, Rockefeller or Kirkpatrick 1.80
California Crab Meat au Gratin en Casserole 1.55
Filet of Turbot a la Bartel 1.50
Chioppino or Bouillabaisse: A Fish Stew made with Clams, Lobster and Crab Meat,
     Halibut, Salmon, and french Croutons, Imported Saffron 1.85

ASK YOUR ATTENDANT FOR WINE AND LIQUOR MENU

Figure 7.2.   Bernstein's Fish Grotto menu, c. 1952.

communities. In 1899 a reporter observed Italian fishermen preparing a hot breakfast on Fisherman's Wharf before going back out on the water: "cold and hungry as they are, the soup or stew must be well cooked before it is eaten."[16] He also noted some of their supplies—flat loaves of Italian bread, baskets of onions, and bottles of wine—but that was enough culinary flavor for his story. When the future aviator Harriet Quimby spent a day with Italian fishermen two years later, she turned up much more:

> Whoever goes out in a fishboat, if they have the good fortune not to be sick, should insist upon having a dish of "Chespini." This is the way to make it. My authority is a scrap of soiled paper written in Italian. Translated, it says: "Put into kettle half glass of sweet oil, one clove of garlic, two large tomatoes, two chili peppers, one glass of white wine; prepare fresh fish, cut in small squares, drop into the sauce and cook three minutes; serve hot." It really tastes much better than it sounds.[17]

Even with the approximate spelling, the dish is unmistakably an early version of cioppino.

When Quimby tasted it in 1901, the dish was only available for fishermen or their occasional guest. But by 1906 the post-earthquake *Refugees' Cook Book* included a recipe for "Chippine," suggesting San Franciscans had started to make the hearty fish stew at home. Before long, Bernstein's Fish Grotto, Di Maggio's Grotto, and other local seafood restaurants all had a version of cioppino to offer the eager crowds. Caen's recommendation: wear a bib.[18]

Crab Louis

Crab Louis, with local Dungeness crab in a spicy cream sauce, has been a popular salad up and down the West Coast since the early twentieth century. Its origins are disputed, but in 1908 the Bergez-Frank's Old Poodle Dog's menu listed "Crab Leg à la Louis (special)" for fifty cents.[19] The restaurant's chef, Louis Coutard, apparently made a specialty of serving crab with chili sauce. Coutard had long been the chef at Frank's Rotisserie. After the 1906 earthquake, Frank's Rotisserie joined with Bergez's Restaurant and the Old Poodle Dog to form a new enterprise, called the Bergez-Frank's Old Poodle Dog Co. Coutard was briefly a partner in the new restaurant but died in May 1908. His colleagues honored his invention by listing the dish under his name. (Two years later, Victor Hirtzler published a recipe for "Crab à la Louise," made with sweet Spanish pimentos instead of the spicy chili sauce.)[20]

The taste for Crab Louis soon spread all over the city. San Franciscans already associated Dungeness crab with celebrations—crab season usually opened either in early November or early December, so locals traditionally enjoyed crab for the holidays. White-aproned vendors on Fisherman's Wharf boiled fresh crab in large kettles and sold it to passersby: a familiar sight to locals, but picturesque enough for a tourist's postcard. With the creation of Crab Louis, San Franciscans could now spice up their crab and give it some flair.

Chronicle restaurant critic Edwords favored a version prepared by Solari's Grill, on Geary:

> **Crab Louis**: Take meat of crab in large pieces and dress with the following: One-third mayonnaise, two-thirds chili sauce, small quantity chopped English chow-chow, a little Worcestershire sauce and minced tarragon, shallots and sweet parsley. Season with salt and pepper and keep on ice.[21]

The 1915 *Pan-Pacific Cook Book* substituted catsup for both the chili sauce and chowchow (pickled relish), but dressed it up by serving the salad in cocktail glasses.

Changing tastes have showcased other ways of serving crab. Since 1971, for instance, crowds have flocked to Thanh Long, one of San Francisco's oldest Vietnamese restaurants, for its succulent roast crab, and later for drunken crab and tamarind crab, too. But over the years, most San Francisco seafood restaurants have kept Crab Louis on the menu, promoting it as one of the city's historic dishes. Many restaurants also offer the same creamy Louis sauce with its chili kick, but as a dressing for shrimp or lobster salads as well as the traditional crab.

Hangtown Fry

At the start of the Gold Rush, miners were eager to demonstrate their wealth by conspicuously consuming expensive meals. Herds of cattle on the ranchos meant that big steaks were not hard to procure; chickens, on the other hand, proved much harder to raise successfully. Seabird eggs from the Farallon Islands often substituted for the preferred kind. The desire for chicken eggs was so intense that clipper ships brought four-month-old eggs from Boston. Meanwhile, fresh eggs cost as much as two or three dollars each in mining country. In later years, forty-niners remembered those early years when they survived on beans, bacon, and coffee, with eggs as a rare treat. Canned oysters were common in mining country, but fresh oysters were about as scarce as eggs.[22]

Shortly after gold was discovered at Coloma, miners founded a camp nearby called Dry Diggings. In early 1849 a French miner claimed two

Mexicans robbed him of his gold, and others accused the same men of being horse thieves. A mob hung the accused from an oak tree in the center of the settlement, which then acquired the name Hangtown. Community leaders soon renamed the town Placerville, but the colorful nickname stuck. In 1850 the El Dorado Hotel and Saloon opened on that same square, next to the infamous oak tree. The hotel restaurant gained fame for its excellent food, and in later years many authors reprinted the El Dorado's bill of fare, advertising roast beef, roast grizzly bear, bean soup, sauerkraut, fried bacon, and "18-carat hash," each dish costing one dollar. The cheapest item was two medium-sized potatoes for fifty cents; diners could enjoy ordinary hash or baked beans for seventy-five cents. The menu specified that diners had to pay in advance, and a scale was available for weighing any payment in gold. Neither eggs nor oysters appeared on the menu, but an egg might be scrambled into a pan of "18-carat hash" to justify the name.[23]

Legends say one day some happy miner at the El Dorado demanded the most expensive dish possible. The cook satisfied him with a scramble of eggs and oysters. (Some stories add bacon as well.) More likely, miners arrived in San Francisco expecting high prices for dishes with eggs and oysters. Although the price of eggs began declining, the price of oysters was rising over time. San Francisco chop houses and saloons were happy to charge a premium for oyster omelets or scrambles. In 1919 Victor Hirtzler finally gave the dish official recognition, publishing the following recipe:

> **Hangtown Fry**: Mix plain scrambled eggs with one dozen small fried California oysters.[24]

It may have been hard for readers around the country to understand the fuss over this simple dish. For San Franciscans, Hangtown Fry has long provided a way to imagine the forty-niners' experience. The high cost of living in the Bay Area becomes easier to bear when considering miners paying out their hard-won gold for an overpriced oyster omelet.

Mixed Drinks

Pisco Punch: From the Gold Rush days, San Franciscans enjoyed Peruvian "pisco," white grape brandy imported on ships traveling up the western coast of South America after rounding Cape Horn. Pisco had a harsh edge, suitable for the frontier town. As the city started to become more sophisticated in tone in the 1880s, Duncan Nicol at the Bank Exchange Bar popularized a much smoother cocktail named "Pisco Punch." He gave his drink an aura of exclusivity by maintaining complete secrecy about the recipe. Even after Prohibition shut down the Bank Exchange, he would not reveal the recipe: "Mr. Volstead can't take the secret from me," Nicol declared.[25] Imbibers could taste simple syrup, lemon juice, and pineapple syrup, and they could see the pineapple chunks; soon other bars served their own version of the famous Pisco Punch. But how Nicol smoothed out the roughness of the brandy remained a mystery until a historian tracked down the missing ingredient in 1973: gum arabic. Gum arabic, made from the sap of the acacia tree, made Pisco Punch go down "as lightly as lemonade," while finishing with "the kick of a roped steer."[26] Nicol's secret recipe brought tourists and locals to the Bank Exchange for decades:

> Ladies . . . used to enter the side portals of the Bank Exchange with a thrill of unholy delight and drink one or even two Pisco Punches. . . . [That] was not regarded as drinking in the common or garden sense of that term, but "seeing life," and was regarded as a rite akin to eating chop suey in Chinatown.[27]

Gold/Silver Fizz: Other mixed drinks also took hold in fin-de-siècle San Francisco. One hotel bartender described the drinks he was regularly asked to make. Among the local specialties he mentioned were a "Gold Fizz" and a "Silver Fizz," each made with gin, lemon juice, sugar, and seltzer, but with an egg yolk mixed into the Gold Fizz and an egg white in the Silver. Men ordered these in the morning, to start the day off with a nutritious kick; the drinks also served to celebrate the gold and silver bonanzas then coming out of the Comstock Lode.[28]

Martini: In the 1890s San Franciscans started enjoying martinis, although what they drank bore little resemblance to the much drier martini of the 1950s. Sometimes called the Martinez, this drink of sweet gin and sweet vermouth was a variation on a Manhattan, sometimes to the point of being garnished with a maraschino cherry rather than an olive. Women liked these martinis as much as they did Pisco Punch, while men spoke out against the sweet version: "The dry Martini, with a stuffed California olive reposing in the bottom of the glass, was calculated to assure a measure of respect at all times."[29] (Local fruit growers approved of both choices—olive or cherry.) In later years, Herb Caen mocked the old Martinez but also poked fun at those obsessed with the driest of dry martinis:

> Customer: "Gimme a mental martini."
> Bartender, "Whazzat?"
> Customer: "You pour the gin, I'll think vermouth," which, as martini jokes go, is fairly dry.[30]

Irish Coffee: Early San Franciscans had an affinity for "Coffee Royals," strong coffee mixed with Irish whiskey. But only in the 1950s did locals begin to associate San Francisco's foggy climate with the need for a sweet warm alcoholic beverage. In 1952 *Chronicle* travel writer Stanton Delaplane came back from Europe nostalgic for a coffee drink he had enjoyed at Ireland's Shannon Airport. He teamed up with the owner of the Buena Vista Saloon, Jack Koeppler, and experimented until they came up with the rich, velvety Irish coffee. There were some problems along the way: they wanted to top it with whipped cream, but at first the cream tended to sink into the whiskey and coffee. Dairyman (and San Francisco mayor) George Christopher figured out the solution: let the cream sit two days before frothing it to the right consistency. In 1953 Delaplane's column went into syndication, giving him a national platform to promote the Buena Vista's popular new drink. A little attention from Herb Caen cemented Irish coffee's status as a San Francisco tradition.

Mai Tai: In 1944 Trader Vic poured a shot of seventeen-year-old Jamaican rum into a shaker, squeezed in some fresh lime, then added orange Curaçao, rock candy syrup, just a bit of almond liqueur, and quite a lot of shaved ice. After a vigorous shaking, he offered the new drink to some friends visiting from Tahiti. Their response: "*Mai Tai-Roa Ae*" (Tahitian for "the best!"). For Vic, that was destiny. Both the recipe and the name were set; the drink's fame soon spread far beyond the Bay Area. A half-century later, mixologist Tony Abou-Ganim was inspired by Trader Vic to come up with his own San Francisco landmark drink. In Harry Denton's Starlight Room, at the top of the Sir Francis Drake Hotel, Abou-Ganim frosted a chilled glass with cinnamon-sugar. He combined spiced rum, orange Curaçao, lemon juice, and simple syrup, and considered names for his new drink. Up there one felt halfway "between the stars and the cable cars." A new classic was born: the Cable Car.[31]

Mission Burrito

Imagine a tube the size of a bazooka shell, filled with spicy grilled meat, guacamole, salsa, tomatoes, refried beans, rice, onions and cilantro.

[It's] like eating a living, breathing organism—you can feel the burrito's ingredients sigh inside with each bite, each squeeze.[32]

Sometimes called the "Silver Torpedo" for the tinfoil keeping the massive contents from bursting out, the Mission burrito is a startling contemporary of the California Cuisine movement. As Berkeley gourmets were learning to spend more for cheese, coffee, meat, and produce, Mission burritos offered a full meal for less than two dollars. The cost of living in San Francisco kept going up, and yet students and low-paid workers survived, fueled by these huge burritos.

Burritos did not start out enormous. At the turn of the twentieth century, "burrito" was a regional term for food rolled in a tortilla, much like a soft taco. In mid-century Northern California, most people eating burritos were farmworkers who needed a hearty breakfast before the heat of the day. Peter Garin, now a San Francisco restaurant consultant, picked lettuce in the Salinas Valley in the 1960s. He remembered looking forward to the "full spicy killer burritos at around 10:30, keep you going till afternoon."[33]

El Faro, at 20th and Folsom, claims it produced the very first extra-large Mission burrito. On September 25, 1961, neighborhood firemen came into Febronio Ontiveros's new grocery store, El Faro (The Lighthouse), looking for sandwiches. The next day, Ontiveros prepared overstuffed burritos for the firemen by overlapping three ordinary wheat tortillas.

The new Mission burrito's popularity led to larger tortillas—always wheat, as corn tortillas would fall apart at that size. Another Mission district taqueria, La Cumbre, came up with key innovations, including assembly-line production allowing diners to select their preferred beans, meat, and degree of spiciness. In 1979 San Francisco food columnist Patricia Unterman helped teach San Franciscans about Mission burritos. She raved about the juicy burritos at Panchito's Tijuana Village, with their charcoal-grilled meats, but she also had a soft spot for Taqueria Tepatitlan's burritos: "the ones with creamy guacamole practically flow out of their tortilla, like melted ice cream out of the cone."[34] John Roemer observed, in a famous 1993 *SF Weekly* article on the "Cylindrical God," that San Franciscans took to the hearty sustenance of the Mission burrito as if it completed their souls, transcending the physical hunger of the body.

Rice-A-Roni

In the 1890s, Charlie DeDomenico immigrated to San Francisco. After the great earthquake, he arranged for the Ferrignos, a pasta-producing family, to join him from Naples, Italy. Marrying Maria Ferrigno, he began an empire that eventually earned more than $250 million in annual sales and wedded the company's most famous product to San Francisco's image. By 1942, his company was called the Golden Grain Macaroni Company. The patriarch died the next year, but his sons had the company well in hand.

In 1946 the youngest son, Tom DeDomenico, married a young immigrant, Lois Bruce, and for a few months they rented a room from an old Armenian widow, Pailadzo Captanian. Lois was pregnant, and Canadian. Tom's mother began teaching her to cook Italian food, but Lois also spent time with Mrs. Captanian in the kitchen, learning to make baklava and Armenian pilaf. The widow also talked about her sad experiences during the Armenian Genocide, adding poignant context to the pilaf lessons.[35]

Tom would bring home Golden Grain vermicelli from the factory, and Lois learned to break it into tiny pieces for the pilaf. When they moved out, Lois kept making pilaf, perhaps simplifying it somewhat: "You sautéed the rice and vermicelli in butter until a light tan, poured over the can of [Swanson's] chicken broth, covered and simmered until the moisture was absorbed."[36] Everyone loved Lois's special dish. Tom's elder brother Vincent particularly noted how easy it was to prepare. The brothers saw they could readily adapt their packaged soup mixes to this new flavor profile; they could then market the dish as a tasty, convenient alternative to potatoes on the dinner plate.

The company already had "macaroni" in its name; Vincent suggested calling the product "Rice-a-Roni," and his brothers agreed it had a nice ring. Vincent may have been inspired by jazz musician Slim Gaillard, who had a club in the Fillmore district and was always saying "Right-orooni" or "How 'bout a little bourbon-orooni."[37] Associating Rice-A-Roni with San Francisco made sense, as the company was founded there and the city was already famous for great cosmopolitan cuisine. In 1958 Golden Grain Macaroni began selling Rice-A-Roni in California, where customers loved it. By 1961 Rice-A-Roni ads were plastered on cable cars, and those cable cars then starred in catchy television commercials, distributed nationally on ABC. Golden Grain Macaroni maintained the cheerful advertising campaign over decades, providing free publicity for San Francisco. It was a happy union: when local financier Richard Blum married then-mayor Dianne Feinstein in 1980, guests received adorable pouches of Rice-A-Roni instead of rice to throw at the happy couple.

Sourdough Bread

Legends about San Francisco sourdough link the city's beloved bread to Gold Rush miners, rough-and-ready bakers far from civilization. To be clear, nothing is uncivilized about making bread with naturally available yeasts and bacteria. For thousands of years, bakers around the world have allowed "wild" yeasts and lactobacilli to colonize their flour, helping it rise. From very early on, bakers learned to save time by using a starter, a small amount of old dough reserved from a prior batch. They also began to simplify the process by using brewers' yeast, and later bakers' yeast.

By the early nineteenth century, other commercial products emerged to simplify and standardize bread baking. In 1846 New Englanders John Dwight and Austin Church began producing baking soda made from sodium carbonate and carbon dioxide. During the California gold rush, the forty-niners were delighted to be able to use baking soda to speed up the rising process. They were busy mining for gold and did not have time to spend making bread. Trading posts kept miners well connected to commerce and civilization. Here is how one California miner baked his bread:

> Taking a tin pan, which served alternately as a gold-washer and a bread-tray, I turned into it a few pounds of flour, a small solution of saleratus [baking soda], and a few quarts of water, and then went to work in it with my hands, mixing it up and adding flour till I got it to the right consistency; then shaping it into a loaf, raked open the embers, and rolled it in, covering it with the live coals. . . . In half an hour or so my bread was baked . . . a little burnt on one side, and somewhat puffed up, like the expectations of the gold-digger in the morning.[38]

Commercial advertisements taught new miners how to use baking soda in their baking.[39]

Some miners did use the older baking methods, but they were up in Alaska. During the 1896–1899 Klondike Gold Rush, miners found themselves isolated for months at a time. They survived by letting flour ferment or "sour" over time, or by preserving starters in special pots worn on the body to keep the starter from freezing. These Alaskan miners earned the nickname "sour doughs." San Franciscans welcomed them warmly for the Mining Fair, part of the city's Golden Jubilee.

The 1898 Mining Fair showcased San Francisco as a jumping-off point for miners headed to the Klondike and encouraged investment in California's own mining industry. The Fair was held at the Mechanics' Pavilion—now the site of the Civic Auditorium. In the "Klondike Kitchen," fairgoers learned how to make sourdough bread and sourdough cornbread, as well as desiccated potato soup and broiled herrings.[40] The swirl of excitement associating mining with San Francisco and the Jubilee must have led many to confuse the Klondike "sour doughs" with the famous forty-niners. After 1906 San Francisco proved happy to take advantage of that confusion in its promotional efforts.

During the nineteenth century, writers did not describe San Francisco's bread as having a distinctive sour taste. Bakers using French sourdough (*pain de levain*) techniques took care to avoid sour-tasting bread.[41] Only in the twentieth century did sourdough bread begin to be praised for its tang. In 1904 an unnamed expert—"who knows more about bread than any other man in San Francisco"—spoke highly of the city's sourdough:

French bread is entirely different from Italian bread. . . . Both, however, are sourdough breads and should taste sour. . . . Although the French and Italian loaves look much alike, the Frenchman works his bread quite a bit more and makes a springier loaf. . . . These French loaves are very tough, are filled with large holes, and are, so far as I can find, confined to San Francisco. I've offered 50 cents for a loaf of it in New York, and it can't be got. It's not made in Seattle, in Portland, nor in Tacoma. There may be a stray baker of it in Los Angeles or Sacramento, but that's all. It's San Francisco and only San Francisco.[42]

This was the first mention of sourdough bread as a San Francisco specialty. At the city's 1915 Panama-Pacific International Exposition, the Sperry Flour Company's baking exhibit included a former Klondike miner making sourdough, again associating that bread both with mining and with San Francisco, as at the 1898 Mining Fair. This time, however, those sampling the sourdough were not potential miners but tourists and appreciative San Franciscans.[43] First prize in bread making went to a San Francisco bakery founded by Jean Larraburu, a Basque immigrant. The Larraburu Brothers Bakery became known for its distinctive sourdough bread, enjoyed by San Franciscans for seventy years.

The Boudin bakery has San Francisco roots at least as far back as 1852. Isidore Boudin, who had trained as an apprentice *boulanger* in France, used French methods in his establishment on Dupont Street. After 1887, matriarch Louise Boudin ran the business, moving it to Broadway in 1895. When the 1906 earthquake hit, the story is that Louise grabbed a bucket of the original starter before running to safety. She instinctively protected the "mother dough," which linked Boudin's bread back to its beginnings. Boudin Bakery soon reopened on Tenth Avenue, where it is today, still baking bread whose story goes back to the very birth of San Francisco.[44]

Other old established houses such as the Parisian Bakery gradually began advertising sourdough breads. By the 1940s, San Francisco bakeries sold both French loaves and sourdough loaves—intensifying the tang in the latter as people began to relish it more.[45] In 1971 scientists identified the bacteria responsible for the bread's sour flavor, naming it after San Francisco even though the same bacteria exist around the world. In the 1980s artisanal companies including Acme Bread and Semifreddi's sprang up, reinforcing the region's reputation for delicious sourdough bread and then spreading the taste for sourdough around the world. Tourists and locals all grew to love the idea of San Francisco's fog and salt air giving a special sour flavor to the city's bread.[46] As with other beloved San Francisco treats, the city's affection for a distinctive taste entwined sourdough bread into San Francisco's storied past.

~

Notes

Introduction

1. *SFC* (Dec 12, 1891), 8.

2. *Conversations With M. F. K. Fisher*, ed. David Lazar (Jackson: University Press of Mississippi, 1992), 104.

Chapter One: The Material Resources

1. Barbara L. Voss, *The Archaeology of Ethnogenesis: Race and Sexuality in Colonial San Francisco* (Berkeley: University of California Press, 2008), 60–64, 83–89, 216–44; M. L. A. Milet-Mureau, *Voyage de la Pérouse autour du monde*, vol. 2 (Paris: Plassan, 1797), 266.

2. Milet-Mureau, *Voyage*, 266. Maria F. Wade, *Missions, Missionaries, and Native Americans: Long-Term Processes and Daily Practices* (Gainesville: University Press of Florida, 2008), 195–200, 208; Quincy D. Newell, *Constructing Lives at Mission San Francisco: Native Californians and Hispanic Colonists, 1776–1821* (Albuquerque: University of New Mexico, 2009), 56–58, 79.

3. Joseph R. Conlin, *Bacon, Beans, and Galantines: Food and Foodways on the Western Mining Frontier* (Reno: University of Nevada Press, 1987), 118–20, 127.

4. William Shaw, *Golden Dreams and Waking Realities: Being the Adventures of a Gold-Seeker in California* (London: Smith, Elder, 1851), 39–42. See also Conlin, *Bacon, Beans, and Galantines*, 88–95, 109, 176–90.

5. Bayard Taylor, *Eldorado, or, Adventures in the Path of Empire: Comprising a Voyage to California, Via Panama*, vol. 1 (New York: Henry G. Bohn, 1850), 87–88.

6. Richard A. Walker, *The Conquest of Bread: 150 Years of Agribusiness in California* (New York: The New Press, 2004), 40–42, 240–42.

7. Ernest Peixotto, "Italy in California," *Scribner's Magazine*, vol. 48 (1910), 78; Simone Cinotto, *Soft Soil, Black Grapes: The Birth of Italian Winemaking in California* (New York: New York University Press, 2012), 132–33.

8. Agoston Haraszthy, "The Early History of the Vine-Culture in California," *Transactions of the California State Agricultural Society* (1859), 312.

9. SFC (May 24, 1878), 3; *The State* (June 7, 1879); "The Water-Wagon," *San Francisco Water* (July 1922), 12.

10. Norris Hundley Jr., *The Great Thirst: Californians and Water—A History* (Berkeley and Los Angeles: University of California Press, 2001), 193.

11. *Roughing It* (New York: Harper & Brothers, 1913), 140; *The Daily Morning Chronicle* (Oct. 22, 1868), 1; *The Daily Morning Chronicle* (Oct. 24, 1868), 3.

12. *Appendix to the Journals of the Senate and Assembly of the Legislature of the State of California*, vol. 2 (1907), 14.

13. Marshall Everett, *Complete Story of the San Francisco Earthquake* (Chicago: Bible House, 1906), 161–65, 196–200.

14. Charles Morris, *The San Francisco Calamity by Earthquake and Fire* (Philadelphia: J. C. Winston, 1906), 132.

15. SFC (Apr. 27, 1906), 9; SFC (Apr. 30, 1906), 12.

16. SFC (Apr. 30, 1906), 4; SFC (July 8, 1906), 29; SFC (July 24, 1906), 12; SFC (Nov. 18, 1906), B60.

17. *The Economist* (Oct. 21, 1989), 26.

18. Virginia de Araujo, "Loma Prieta," *San José Studies*, vol. 16 (San Jose: California State University Press, 1990), 12.

Chapter Two: Native American Foodways

1. Alfred Louis Kroeber, *Shoshonean Dialects of California* 4:1–6 (Berkeley: University of California Press, 1907), 200–202.

2. Randall Milliken et al., "Punctuated Culture Change in the San Francisco Bay Area," in *California Prehistory: Colonization, Culture, and Complexity*, ed. Terry L. Jones and Kathryn A. Klar (Lanham, MD: AltaMira, 2010), 114.

3. Milliken, "Punctuated Culture Change," 106.

4. Kent Lightfoot and Otis Parrish, *California Indians and their Environment* (Berkeley: University of California Press, 2009), 124–26, 134–40; M. Kat Anderson, *Tending the Wild: Native American Knowledge And the Management of California's Natural Resources* (Berkeley: University of California Press, 2006), 182–87.

5. Sir Francis Drake, "The Voyage About the World," in *The English Circumnavigators*, ed. David Laing Purves (London: William P. Nimmo, 1874), 79.

6. Alfred Louis Kroeber et al., *A Mission Record of the California Indians*, vol. 8 (Berkeley: The University Press, 1910), 22.

7. Bertha H. Smith, "The Indian Breadmaker," *Good Housekeeping* 39 (1904), 189.

8. "Indianology," *California Farmer and Journal of Useful Sciences* 12:9 (Apr. 20, 1860), 66; Mary J. Gates, *Contributions to Local History: Rancho Pastoria de los Borregas, Mountain View, California* (San Jose: Cottle & Murgotten, 1895), 14–15; Mary Sheldon Barnes, "Some Primitive Californians," *The Popular Science Monthly* 50 (1897), 494; Genevra Sisson Snedden, *Docas, the Indian Boy of Santa Clara* (Boston: D. C. Heath, 1899), 6–7; Margaret Denise Dubin and Sara-Larus Tolley, *Seaweed, Salmon, and Manzanita Cider: A California Indian Feast* (Berkeley: Heyday Books, 2008), 103.

9. Francisco Palóu, *The Life and Apostolic Labors of the Venerable Father Junipero Serra*, vol. 1, trans. C. Scott Williams (Pasadena: G. W. James, 1913), 209–14.

10. Jean-François de Lapérouse, *Voyage de La Pérouse autour du monde*, vol. 2 (Paris: Plassan, 1798), 299.

11. *Mémoires du Capitaine Péron Sur ses Voyages*, vol. 2 (Paris: Bossange Frère, 1824), 128–29.

12. Lightfoot and Parrish, *California Indians*, 217.

13. Malcolm Margolin, *The Ohlone Way: Indian Life in the San Francisco-Monterey Bay Area* (Berkeley: Heyday Books, 1978), 19, 48–50; C. Hart Merriam, "The Acorn, A Possibly Neglected Source of Food" (1918), in *Food in California Indian Culture*, ed. Ira Jacknis (Berkeley: University of California Press, 2004), 149; Lightfoot and Parrish, *California Indians*, 130.

14. Kroeber, *Mission Record*, 22.

15. Lightfoot and Parrish, *California Indians*, 222–26; Ira Jacknis, "Introduction," *Food in California Indian Culture*, (Berkeley: University of California, 2004), 46.

16. Snedden, *Docas*, 44–45; Margolin, *The Ohlone Way*, 25–35, 81.

17. Snedden, *Docas*, 16.

18. Mabel Miller, "The So-Called California 'Diggers,'" *Popular Science* (December 1896), 205; Margolin, *The Ohlone Way*, 41–45; Lightfoot and Parrish, *California Indians*, 218.

19. Pedro Fages, *Expedition to San Francisco Bay in 1770*, ed. Herbert Eugene Bolton (Berkeley: University of California Press, 1911), 12–13.

20. Lightfoot and Parrish, *California Indians*, 220–44; Lowell John Bean, *The Ohlone, Past and Present* (Menlo Park: Ballena Press, 1994), 48, 263.

21. Margolin, *The Ohlone Way*, 7–9.

22. Les W. Field, *Abalone Tales: Collaborative Explorations of Sovereignty and Identity in Native California* (Durham: Duke University Press, 2008), 150, 175.

23. Lightfoot and Parrish, *California Indians*, 24–28, 216, 236–37.

24. Randall Milliken, *A Time of Little Choice: The Disintegration of Tribal Culture in the San Francisco Bay Area, 1769–1810* (Menlo Park: Ballena Press, 1995), 73.

25. Sherburne Friend Cook, *The Conflict Between the California Indian and White Civilization* (Berkeley: University of California Press, 1943), 100, 236, 463, 491–94.

26. Cook, *The Conflict*, 287–89.

27. Sir George Simpson, *An Overland Journey Round the World: During the Years 1841 and 1842* (Philadelphia: Lea and Blanchard, 1847), 177.

28. Robert F. Heizer and Alan J. Almquist, *The Other Californians: Prejudice and Discrimination Under Spain, Mexico, and the United States to 1920* (Berkeley: University of California Press, 1977), 20.

29. *DAC* (July 29, 1864), 1.

30. Kurt M. Peters, "Boxcar Babies: The Santa Fe Railroad Indian Village at Richmond, California, 1940–1945," in *Native American Perspectives on Literature and History*, ed. Alan R. Velie (Norman: University of Oklahoma Press, 1994), 100.

31. Wendy Rose, "To the Hopi in Richmond (Santa Fe Indian Village)" (1983), in *Heath Anthology of American Literature*, ed. Paul Lauter and Juan Bruce-Novoa (Lexington, MA: D. C. Heath, 1990), 2528.

32. Adam Fortunate Eagle and Tim Findley, *Heart of the Rock: The Indian Invasion of Alcatraz* (Norman: University of Oklahoma Press, 2002), 32.

33. Susan Lobo, *Urban Voices: The Bay Area American Indian Community* (Tucson: University of Arizona Press, 2002), 52.

34. Fortunate Eagle and Findley, *Heart of the Rock*, 131–38.

35. Recipe adapted from Poncho's "Fry Bread Recipe," in Lobo, *Urban Voices*, 52.

Chapter Three: Immigrants and Ethnic Neighborhoods

1. Jacqueline Higuera McMahan, *Rancho Cooking: Mexican and Californian Recipes* (Seattle: Sasquatch Books, 2001).

2. *SFC* (May 12, 1890), 3.

3. *SFC* (Sept. 2, 1893), 4.

4. Hubert Howe Bancroft, *History of California*, vol. 21 (San Francisco: The History Company, 1886), 175.

5. David E. Hayes-Bautista, *El Cinco De Mayo: An American Tradition* (Berkeley: University of California Press, 2012), 4, 126–27.

6. *SFC* (June 22, 1884), 1.

7. Gustavo Arellano, *Taco USA: How Mexican Food Conquered America* (New York: Scribner, 2012), 38–46; *SFC* (July 14, 1895), 10; *SFC* (Jan. 29, 1905), 7.

8. *SFC* (Jan. 4, 1884), 4.

9. *SFC* (Jan. 2, 1887), 1; *SFC* (Oct. 21, 1906), 2; *SFC* (July 21, 1907), A8.

10. *SFC* (June 16, 1889), 1.

11. *SFC* (July 20, 1890), 6; *SFC* (Aug. 16, 1896), 9.

12. David R. Diaz, *Barrio Urbanism: Chicanos, Planning, and American Cities* (New York: Routledge, 2005), 37, 179.

13. John Krich, "San Francisco's Real Mission," *NYT: The Sophisticated Traveler Magazine* (Oct. 1, 1989), 28.

14. Hubert Howe Bancroft, *History of California*, vol. 6 (San Francisco: The History Company, 1888), 190.

15. *SF-Call* (Jan. 12, 1904), 16; Charles C. Dobie, *San Francisco: A Pageant* (New York, Appleton-Century, 1933), 317.

16. Jessica B. Harris, *High on the Hog: A Culinary Journey from Africa to America* (New York: Bloomsbury, 2011), 151–52; Mary Ellen Pleasant Papers, Folder 21 of Helen Holdredge Collection, San Francisco Public Library History Center.

17. *DAC* (Apr. 7, 1854), 2; Quintard Taylor, *In Search of the Racial Frontier: African Americans in the American West, 1528–1990* (New York: Norton, 1998), 83.

18. *SFC* (Jan. 15, 1919), 52.

19. Karen Hess, "What We Know About Mrs. Abby Fisher and Her Cooking," in *What Mrs. Fisher Knows About Southern Cooking* (Bedford, MA: Applewood Books, 1995), 75–90; Doris Witt, "From Fiction to Foodways: Working at the Intersections of African American Literary and Culinary Studies," in *African American Foodways: Explorations of History and Culture*, ed. Anne L. Bower (Chicago: University of Illinois Press, 2007), 107.

20. "The Palace Hotel," *The Overland Monthly* 15 (September 1875), 299; Taylor, *In Search of the Racial Frontier*, 198–99.

21. *DAC* (Nov. 9, 1889), 1.

22. Delilah Leontium Beasley, *The Negro Trail Blazers of California* (Los Angeles: no publisher, 1919), 149–50.

23. *SF-Call* (Sept. 4, 1910), 34; Gretchen Lemke-Santangelo, *Abiding Courage: African American Migrant Women and the East Bay Community* (Chapel Hill: UNC Press, 1996), 73.

24. Thomas Tramble and Wilma Tramble, *The Pullman Porters and West Oakland* (San Francisco: Arcadia, 2007), 47, 83.

25. Mary Praetzellis and Adrian Praetzellis, "'Black is Beautiful': From Porters to Panthers in West Oakland," report by the Anthropological Studies Center at Sonoma State University (June 2004), 280–81, accessed at http://www.sonoma.edu/asc/cypress/finalreport/Chapter10.pdf.

26. Praetzellis and Praetzellis, "'Black is Beautiful,'" 288.

27. Praetzellis and Praetzellis, "'Black is Beautiful,'" 290; Lemke-Santangelo, *Abiding Courage*, 2–3.

28. Shirley Ann Wilson Moore, "'Your Life Is Really Not Just Your Own': African American Women in Twentieth Century California," in *Seeking El Dorado: African Americans in California*, ed. Lawrence B. De Graaf et al. (Los Angeles: Autry Museum of Western Heritage, 2001), 227.

29. Shirley Ann Wilson Moore, "Traditions from Home: African Americans in Wartime Richmond, California," in *The War in American Culture: Society and Consciousness During World War II*, ed. Lewis A. Erenberg and Susan E. Hirsch (Chicago: University of Chicago Press, 1996), 271–75.

30. Lemke-Santangelo, *Abiding Courage*, 72, 76, 85, 90.

31. Elizabeth Pepin and Lewis Watts, *Harlem of the West: The San Francisco Fillmore Jazz Era* (San Francisco: Chronicle Books, 2006), 51.

32. Maya Angelou, *I Know Why the Caged Bird Sings* (Random House Digital, Inc., 2002), 27.

33. Interviews with Willie Brown and Carol O'Gilvie, in "The Fillmore" (PBS documentary and companion website). Accessed June 8, 2012, http://www.pbs.org/kqed/fillmore/learning/people/index.html.

34. Bob Hayes, *The Black American Travel Guide* (San Francisco: Straight Arrow Books, 1971), 254–57.

35. *SFC* (July 13, 1889), 8; Claudine Chalmers, *French San Francisco* (San Francisco: Arcadia, 2007).

36. *SF-Call* (Mar. 5, 1895), 6.

37. *DAC* (May 9, 1853), 1.

38. Joseph Robert Conlin, *Bacon, Beans, and Galantines: Food and Foodways on the Western Mining Frontier* (Reno: University of Nevada Press, 1987), 112.

39. Daniel Lévy, *Les Français en Californie* (San Francisco: Grégoire, Tauzy, 1885), 251.

40. Édouard Auger, *Voyage en Californie* (Paris: Hachette, 1854), 187.

41. M. Derbec, "Lettre écrite de la Californie," *Nouvelles annales des voyages*, vol. 129 (1851), 240.

42. Alexandre Lambert de Sainte-Croix, *De Paris à San Francisco* (Paris: Calmann Lévy, 1885), 191.

43. Pierre Charles Fournier de Saint-Amant, *Voyages en Californie et dans l'Oregon* (Paris: L. Maison, 1854), 453–54.

44. *SFC* (Dec. 26, 1869), 8.

45. Letter by Mr. Mullet, *The Pioneer: California Monthly Magazine* (October 1854), 254.

46. *SFC* (Dec. 26, 1869), 8.

47. "Restaurant," *Dictionnaire de la conversation et de la lecture*, vol. 15, ed. William Duckett (Paris: Michel Lévy, 1857), 379; Hubert Howe Bancroft, "Mongolianism in America," in *Essays and Miscellany* (San Francisco: The History Company, 1890), 331.

48. Kevin Starr and Richard J. Orsi, *Rooted in Barbarous Soil: People, Culture, and Community in Gold Rush California* (Berkeley: University of California Press, 2000), 152–53.

49. *DAC* (Feb. 17, 1852), 2.

50. Conlin, *Bacon, Beans, and Galantines*, 189–92.

51. Albert S. Evans, "From the Orient Direct," *The Atlantic Monthly*, vol. 24 (November 1869), 548.

52. Bancroft, "Mongolianism," 321, 329; Andrew Coe, *Chop Suey: A Cultural History of Chinese Food in the United States* (New York: Oxford University Press, 2009), 129.

53. *SFC* (Apr. 2, 1876), 1; *SFC* (Nov. 19, 1899), 2.

54. Bancroft, "Mongolianism," 318–19.

55. Bancroft, "Mongolianism," 321–22, 330–34.

56. Coe, *Chop Suey*, 176–79; Brian J. Godfrey, "New Urban Ethnic Landscapes," in *Contemporary Ethnic Geographies in America*, ed. Ines M. Miyares and Christopher A. Airriess (New York: Rowman & Littlefield, 2007), 338–40; Thy Tran, "The Reinvention of San Francisco Chinatown Post-1906," paper presented at the Association for the Study of Food and Society Annual Meeting in New Orleans, LA, on June 6, 2008.

57. Rose Hum Lee, *The Chinese in the United States of America* (Hong Kong: Hong Kong University Press, 1960), 261–68; J. A. G. Roberts, *China to Chinatown: Chinese Food in the West* (London: Reaktion Books, 2002), 149–52.

58. *SFC* (Mar. 14, 1891), 7; Herbert Buell Johnson, *Discrimination Against the Japanese in California: A Review of the Real Situation* (Berkeley: Courier Pub. Co., 1907), 27, 101.

59. *SFC* (Aug. 18, 1893), 4; *SFC* (July 3, 1905), 7; *SFC* (May 22, 1905), 14.

60. Michel S. Laguerre, *The Global Ethnopolis: Chinatown, Japantown and Manilatown in American Society* (New York: St. Martin's Press, 2000), 62–67; Johnson, *Discrimination*, 100–104, 125–27.

61. *SFC* (Oct. 21, 1906), 1.

62. Clarence E. Edwords, *Bohemian San Francisco, Its Restaurants and Their Most Famous Recipes* (San Francisco: P. Elder, 1914), 58–59.

63. Sandra C. Taylor, *Jewel of the Desert: Japanese American Internment at Topaz* (Berkeley: University of California Press, 1993), 32–35, 64, 90; Hatsuro Aizawa, et al., *San Francisco's Japantown* (San Francisco: Arcadia, 2005), 7.

64. Dino Cinel, *From Italy to San Francisco: The Immigrant Experience* (Stanford: Stanford University Press, 1982), 31–32.

65. *Report of the Industrial Exhibition of the Mechanic's Institute of the City of San Francisco* (San Francisco: Franklin, 1858), 78.

66. *SFC* (Nov. 12, 1875), 3.

67. Cinel, *From Italy to San Francisco*, 106, 111.

68. Cinel, *From Italy to San Francisco*, 15; Direzione generale della statistica, *Statistica della emigrazione Italiana* (Italy, 1886), 75–76.

69. Giovanni Vigna dal Ferro, *Un viaggio nel Far West americano* (Bologna: Successori Monti, 1881), 42.

70. Ernest Peixotto, "Italy in California," *Scribner's Magazine* 48 (1910), 81.

71. Hasia Diner, *Hungering for America: Italian, Irish, and Jewish Foodways in the Age of Migration* (Cambridge, MA: Harvard University Press, 2001), 120–31.

72. *SFC* (Mar. 18, 1895), 5; *SFC* (Mar. 18, 1922), 13.

73. *SFC* (Dec 12, 1891), 8.

74. *SFC* (Oct. 30, 1921), S9.

75. *SFC* (Aug. 25, 1898), 13.

76. *SFC* (Aug. 20, 1922), S2.

77. *SFC* (Jan. 15, 1884), 8.

78. *SFC* (Feb. 12, 1899), 7.

79. Steven Friedman, *Golden Memories of the San Francisco Bay Area* (San Francisco: Arcadia, 2000), 47–50.

80. *SFC* (Dec. 8, 1889), 2.

81. "Rebecca Hourwich Reyher, oral history conducted in 1973 by Amelia Fry" and "Ansel Adams, oral history conducted 1972–1975 by Ruth Teiser and Catherine Harroun," Regional Oral History Office, The Bancroft Library, University of California, Berkeley.

82. *DAC* (Dec. 30, 1852), 2.

83. Frances Bransten Rothmann, *The Haas Sisters of Franklin Street* (Berkeley: Judah L. Magnes Museum, 1979), 71.

84. Harriet Lane Levy, *920 O'Farrell Street: A Jewish Girlhood in Old San Francisco* (Berkeley: Heyday Books, 1996 [orig. 1937]), 62, 137–41.

85. Fred Rosenbaum, *Cosmopolitans: A Social and Cultural History of the Jews of the San Francisco Bay Area* (Berkeley: University of California Press, 2009), 198, 204–6.

86. *The Adjuster* 27:1 (June 1903), 10.

87. Jerry Flamm, *Good Life in Hard Times: San Francisco's '20s and '30s* (San Francisco: Chronicle Books, 1977), 82.

88. Lydia B. Zaverukha and Nina Bogdan, *Images of America: Russian San Francisco* (San Francisco: Arcadia, 2010), 8–9, 28, 53–56.

89. "Valentina Alekseevna Vernon, oral history conducted in 1980 by Richard A. Pierce," Regional Oral History Office, The Bancroft Library, University of California, Berkeley, 26, 32–34.

90. Matty Simmons and Don Simmons, *On the House* (New York: Coward-McCann, 1955), 205; *The Rotarian* 70:4 (April 1947), 54.

91. Sandra Rosenzweig, "Going Back for Seconds," *New West Magazine* (Oct. 22, 1979), clxxv.

92. *SFC* (Nov. 30, 1902), A3.

93. Estella Habal, *San Francisco's International Hotel: Mobilizing the Filipino American Community in the Anti-Eviction Movement* (Philadelphia: Temple University Press, 2007), 10–11.

94. *Filipinos in San Francisco* (San Francisco: Arcadia, 2011), 8, 17; Laguerre, *The Global Ethnopolis*, 87.

95. Habal, *San Francisco's International Hotel*, 11.

96. Meñez, *Folklore Communication Among Filipinos in California* (New York: Arno Press, 1980), 38–40.

97. Gonzalez, "Gathering Souls with Food," *Filipino American Faith in Action: Immigration, Religion, and Civic Engagement* (New York: New York University Press, 2009), 82–98.

98. Juanita Tamayo Lott, *Common Destiny: Filipino American Generations* (Lanham, MD: Rowman & Littlefield, 2006), 19, 42; *San Francisco Municipal Record* (January–March 1931), 66.

99. *NYT* (July 18, 2010), A23B (national edition).

100. Edwords, *Bohemian San Francisco*, 10.

101. Edwords, *Bohemian San Francisco*, 21.

102. Edwords, *Bohemian San Francisco*, 24–25; *Handbook for San Francisco: Historical and Descriptive* (San Francisco Chamber of Commerce, 1913), 54; "The Sleepless City: Types on Our Streets After Midnight," *SFC* (July 26, 1891), 8.

103. *SFC* (July 25, 1897), 3.

104. Nan Alamilla Boyd, *Wide-Open Town: A History of Queer San Francisco to 1965* (Berkeley: University of California Press, 2003), 4–5. Deadfalls were saloons

offering only wine or beer, although sometimes they did serve food; melodeons were liquor dens with mechanical music players.

105. *SFC* (June 28, 1912), 7.

106. Edwords, *Bohemian San Francisco*, 9, 110.

107. Temple Bailey, "Hospitality and the Bachelor Girl," *SFC* (Feb. 23, 1913), 14.

108. Boyd, *Wide-Open Town*, 5–9.

109. Boyd, *Wide-Open Town*, 21–24, 180.

110. Damon John Scott, *The City Aroused: Sexual Politics and the Transformation of San Francisco's Urban Landscape, 1943–1964* (PhD diss., University of Texas at Austin, 2008), 252.

111. Erick Lyle, *On the Lower Frequencies: A Secret History of the City* (Berkeley: Counterpoint Press, 2008), 134–36.

112. *Out of Our Kitchen Closets: San Francisco Gay Jewish Cooking* (San Francisco: Congregation Sha'ar Zahav, 1987), 1–3, 6.

113. Jack Kerouac, *On the Road* (New York: Penguin, 1991), 173–74.

114. Charles Perry, *Haight Ashbury: A History* (New York: Wenner Books, 2005), 20, 89, 111.

115. Perry, *Haight Ashbury*, 94. The Diggers named themselves after a group of agrarian nonconformists in early modern England; it was also a term for certain Native Americans in the western United States.

116. Peter Coyote, *Sleeping Where I Fall: A Chronicle* (Berkeley: Counterpoint Press, 1999), 132.

117. Tim Hodgdon, *Manhood in the Age of Aquarius: Masculinity in Two Countercultural Communities, 1965–83* (New York: Columbia University Press, 2008), 72.

118. Susan Jacobs, "A Vegetarian Guide to Dining in San Francisco," *Vegetarian Times* 93 (May 1985), 49–50.

119. Mirjana Blankenship, "The Farm by the Freeway," in *Ten Years that Shook the City: San Francisco, 1968–1978*, ed. Chris Carlsson (San Francisco: City Lights, 2011), 219–31.

120. Pam Peirce, "A Personal History of the San Francisco People's Food System," in *Ten Years that Shook the City: San Francisco, 1968–1978*, ed. Chris Carlsson (San Francisco: City Lights, 2011), 232–34.

Chapter Four: Food Markets and Retailing

1. *SFC* (June 25, 1916), 50.

2. *DAC* (June 18, 1853), 2.

3. *DAC* (Sept. 1, 1854), 10.

4. *SFC* (Dec. 28, 1890), 10.

5. *DAC* (Sept. 1, 1854), 10.

6. *SFC* (July 10, 1892), 12.

7. *SF-Call* (May 21, 1891), 3.

8. *SFC* (Mar. 16, 1879), 1.

9. *The Pacific Historian* 4–5 (1960), 164.

10. Rudolph J. Walther, *A Tour to San Francisco, California and Return* (Philadelphia: Walther Printing House, 1916), 115.

11. *SFC* (Jan. 15, 1919), 49.

12. *SFC* (Mar. 21, 1915), 60.

13. *SFC* (Sept. 30, 1872), 4.

14. *SFC* (Oct. 8, 1904), 2.

15. *SFC* (Feb. 21, 1922), 24.

16. *SFC* (Dec. 23, 1882), 3; *SFC* (Dec. 21, 1916), 5.

17. *SFC* (Feb. 12, 1893), 4; *SF-Call* (Jan. 16, 1897), 9; Mitchell Postel, "A Lost Resource: Shellfish in San Francisco Bay," *California History* 67:1 (March 1988), 30–31.

18. *DAC* (Apr. 17, 1880), 1.

19. H. D. B. Soulé, "Old California Market," *California Historical Society Quarterly* 5:1 (March 1926), 85–86.

20. *SFC* (July 30, 1873), 3; *SFC* (Oct. 12, 1919), E2.

21. *SF-Call* (Jan. 17, 1893), 6.

22. *SFC* (June 19, 1904), 5.

23. *SFC* (Sept. 15, 1879), 3; Alessandro Baccari, *San Francisco's Fisherman's Wharf* (San Francisco: Arcadia, 2006), 7, 50; Richard Dillon, *North Beach: The Italian Heart of San Francisco* (Novato, CA: Presidio, 1985), 86–91; Henry A. Fisk, "The Fishermen of San Francisco Bay: The Monopoly of an Industry by Greeks, Italians and Chinese," *The Commons* 10 (October 1905), 546.

24. Ben Adams, *San Francisco: An Informal Guide* (New York: Hill & Wang, 1961), 44.

25. "History of the Cannery," accessed on Feb. 19, 2013, at http://www.delmontesquare.com/.

26. *SFC* (June 25, 1871), 3; *SFC* (Sept. 15, 1879), 3.

27. *SFC* (Sept. 15, 1879), 3.

28. *SFC* (Dec. 24, 1889), 8; *SFC* (Dec. 24, 1891), 12; *SF-Call* (Dec. 23, 1895), 7; *SFC* (Jan. 24, 1897), 9.

29. *SFC* (Mar. 16, 1894), 9.

30. *SFC* (Apr. 21, 1887), 8.

31. *SFC* (Jan. 29, 1893), 11; *SFC* (Nov. 11, 1900), 26. In 1893 Daverkosen married another butcher, John J. Markham, whom she also outlived.

32. *SFC* (May 25, 1893), 12.

33. *SFC* (Mar. 6, 1904), 7.

34. *SFC* (Apr. 16, 1914), 20.

35. Jessica Ellen Sewell, *Women and the Everyday City: Public Space in San Francisco, 1890–1915* (Minneapolis: University of Minnesota Press, 2011), 54–55, 63.

36. Sewell, *Women and the Everyday City*, 55.

37. *SFC* (Sept. 29, 1889), 2.

38. *SFC* (Dec. 3, 1918), 8.

39. *SFC* (Apr. 28, 1889), 6; *SFC* (July 22, 1893), 7.

40. Sewell, *Women and the Everyday City*, 57, 64.

41. *SFC* (May 14, 1869), 2; *DAC* (Sept. 2, 1870), 1.

42. *Original Thoughts, Essays and Stanzas Written by the Pupils of the San Francisco Public Schools: Prize Compositions and Stanzas on San Francisco's Industries* (San Francisco: L. R. Hare & Co., 1894), 58–59.

43. Leah Garchik, "Notes From a Fallen City," *SFC* (Apr. 13, 1997).

44. *SFC* (Nov. 15, 1896), 25.

45. John Shertzer Hittell, *The Commerce and Industries of the Pacific Coast of North America* (San Francisco: A. L. Bancroft & Co., 1882), 567.

46. G. F. Hanson, M.D., "Some Examples of Drug and Food Adulteration," *American Druggist and Pharmaceutical Record* 29 (1897), 221.

47. *SFC* (July 29, 1906), 21.

48. *SFC* (Nov. 15, 1908), 25.

49. Walther, *A Tour*, 116.

50. *DAC* (Nov. 15, 1886), 4; Lorine Swainston Goodwin, *The Pure Food, Drink, and Drug Crusaders, 1879–1914* (Jefferson, NC: McFarland, 1999).

51. *SFC* (Feb. 10, 1895), 16.

52. *SFC* (Oct. 24, 1895), 6.

53. *SFC* (Oct. 26, 1895), 8.

54. *SFC* (Apr. 29, 1896), 6.

55. *SFC* (Apr. 15, 1915), 6.

56. "San Francisco Under the Sway of the Labor Unions," *The American Monthly Review of Reviews* 29 (1904), 227–28. For cost of living, see *California Bureau of Labor Statistics Biennial Report for 1883–1884* (1884), 135; *United States Bureau of Labor Statistics, Retail Prices and Cost of Living Series #315* (1923), 100, 142; *United States Bureau of Labor Statistics, Retail Prices of Food, 1923–36* (1938), 95, 144.

57. William Issel and Robert W. Cherny, *San Francisco, 1865–1932: Politics, Power, and Urban Development* (Berkeley: University of California Press, 1986), 86–88.

58. *SFC* (Mar. 26, 1908), 5.

59. Dorothy Sue Cobble, *Dishing It Out: Waitresses and Their Unions in the Twentieth Century* (Urbana: University of Illinois Press, 1991), 89.

60. "Julia Gorman Porter, oral history conducted in 1976 by Gabrielle Morris," Regional Oral History Office, The Bancroft Library, University of California, Berkeley.

61. John Kagel, "The Day the City Stopped," *California History* 63:3 (Summer 1984), 215–20.

62. Robert W. Cherny and William Issel, *San Francisco: Presidio, Port, and Pacific Metropolis* (San Francisco: Boyd & Fraser, 1981), 55; Cobble, *Dishing It Out*, 89–92.

63. Marc Levinson, *The Great A&P and the Struggle for Small Business in America* (New York: Hill & Wang, 2011), 125–27.

64. Alfred Yee, *Shopping at Giant Foods: Chinese American Supermarkets in Northern California* (Seattle: University of Washington Press, 2003), 60–69, 92, 164.

65. Joe Kane, "The Supermarket Shuffle," *Mother Jones Magazine* (July 1984), 7.

66. Laresh Krishna Jayasanker, "Sameness in Diversity: Food Culture and Globalization in the San Francisco Bay Area and America, 1965–2005," (PhD diss., University of Texas at Austin, 2008), 30–32, 68, 75–90.

67. Edward F. Adams, "Marketing Perishable Products in San Francisco," *Proceedings of the Thirty-Fifth Fruit-Growers' Convention* (State Commission of Horticulture, California, 1909), 181–82.

68. Adams, "Marketing Perishable Products," 183–87.

69. "The War Against the Grocers Lobby in San Francisco: An Interview with John Brucato," *Farmers Market Outlook* (September-October 1996).

70. *Farmers' Market Resource Kit: A Step Toward Making San Francisco a Market City* (San Francisco: Sage, 2005); Sally K. Fairfax et al., *California Cuisine and Just Food* (Cambridge, MA: MIT Press, 2012), 169–71, 191; Alison Hope Alkon and Kari Marie Norgaard, "Breaking the Food Chains: An Investigation of Food Justice Activism," *Sociological Inquiry* 79:3 (August 2009), 295.

Chapter Five: Famous Restaurants

1. Hubert H. Bancroft, "Mongolianism," in *Essays and Miscellany* (San Francisco: The History Company, 1890), 331.

2. John V. Tadich, "The Jugoslav Colony of San Francisco," and J. L. Kerpan, "John V. Tadich, One of the Leading Jugoslav Pioneers," in *The Slavonic Pioneers of California: Published on Occasion of the Diamond Jubilee, 1857–1932*, ed. Vjekoslav Meler (San Francisco: Slavonic Pioneers of California, 1932), 42, 50–51.

3. John Briscoe, *The Tadich Grill: The Story of San Francisco's Oldest Restaurant, with Recipes* (Berkeley: Ten Speed, 2002), 8–9, 30–36. Bećir was anglicized to Becker (or sometimes Baker) in San Francisco directories.

4. R. W. Apple Jr., "Where Poseidon sets a Bountiful Table," *NYT* (Sept. 15, 2004), F1.

5. Briscoe, *The Tadich Grill*, 8, 65, 85; Herb Caen, *New Guide to San Francisco and the Bay Area* (New York: Doubleday, 1958), 98.

6. Briscoe, *The Tadich Grill*, 102–6.

7. Chloe Schildhause, "The Gatekeepers: Tadich Grill's Paul Lovallo, on Power Lunches, Celebrities" (interview conducted in 2012), accessed Dec. 18, 2012, http://sf.eater.com/places/tadich-grill.

8. *DAC* (Dec. 7, 1858), 2.

9. Many twentieth-century authors date the Poodle Dog's origins to 1849. But not until 1866 does a restaurant named the "Poodle Dog" appear in the historical record; *DAC* (Feb. 22, 1866), 1. A retrospective report places the "original" Poodle Dog on the southwest corner of Washington and Dupont; *DAC* (Feb. 24, 1868), 1.

10. "A Restaurateur Gone," *SFC* (Feb. 18, 1872), 3.

11. *DAC* (Feb. 22, 1866), 1.

12. *The Daily Morning Chronicle* (July 4, 1869), 1.

13. *DAC* (Oct. 2, 1873), 3.

14. *The New City Annual Directory of San Francisco* (Bishop & Co., 1875), 969.

15. Emma Frances Dawson, "A Deadhead," *Overland Monthly* 14:5 (May 1875), 428–38.

16. Aimé Jaÿ, *A travers les États-Unis d'Amérique* (Niort: Clouzot, 1885), 98.

17. "The Tale of a Poodle," brochure issued by the Poodle Dog Restaurant (San Francisco: Louis Roesch, 1903).

18. *Journal of the Pacific Telephone and Telegraph Company* (1952), 28.

19. *SFC* (Dec. 13, 1898), 12.

20. *Handbook for San Francisco* (1913), 52.

21. Clarence Edgar Edwords, *Bohemian San Francisco, Its Restaurants and Their Most Famous Recipes* (San Francisco: P. Elder, 1914), 17.

22. James R. Smith, *San Francisco's Lost Landmarks* (San Francisco: Quill Driver, 2005), 180; Doris Muscatine, *A Cook's Tour of San Francisco* (New York: Scribner, 1963), 63.

23. *The Daily Dramatic Chronicle* (June 15, 1867), 2; Daniel O'Connell, *The Inner Man: Good Things to Eat and Drink and Where to Get Them* (San Francisco: Bancroft, 1891), 74; Deanna Paoli Gumina, "Provincial Italian Cuisines: San Francisco Conserves Italian Heritage," *The Argonaut: Journal of the San Francisco Historical Society* 1:1 (Spring 1990).

24. Benjamin E. Lloyd, *Lights and Shades in San Francisco* (San Francisco: Bancroft, 1876), 64.

25. *SFC* (June 6, 1871), 3.

26. *SFC* (May 8, 1880), 1; *SFC* (May 23, 1880), 5; *SFC* (May 29, 1880), 3.

27. "Julius Wangenheim: An Autobiography," *California Historical Society Quarterly* 35:2 (1956), 140.

28. Caroline Wells Healey Dall, *My First Holiday: Letters Home from Colorado, Utah, and California* (Boston: Roberts Bros., 1881), 192–95; Bailey Millard, "San Francisco in Fiction," *The Bookman* 31 (1910), 596.

29. Gumina, "Provincial Italian Cuisines"; Edwords, *Bohemian San Francisco*, 25; Ruth Thompson, *Eating around San Francisco* (1937), 103. Campagnoli and her husband, Armando, later opened another restaurant at 869 Geary.

30. "A Trip to the Cliff House," *San Francisco Daily Morning Call* (June 25, 1864).

31. "Early Rising As Regards Excursions to the Cliff House," *The Golden Era* (July 3, 1864), 4.

32. *DAC* (Aug. 8, 1869), 3.

33. *DAC* (Mar. 2, 1871), 2.

34. Edward W. Townsend, "Adolph Sutro: Mayor-Elect of San Francisco, A Capitalist's Fight Against Monopoly," *Review of Reviews* 10 (1894), 628.

35. *SFC* (May 14, 1903), 7.

36. *SFC* (Dec. 4, 1920), 7.

37. Smith, *Lost Landmarks*, 65.

38. Herb Caen, *Only in San Francisco* (New York: Doubleday, 1960), 166.

39. Menu accessed Nov. 30, 2012, at http://thepalacehotel.org.

40. *SFC* (Sept. 18, 1887), 13.

41. *SFC* (May 2, 1891), 3.

42. *SFC* (Dec. 16, 1909), 11.

43. *SFC* (Dec. 29, 1895), 27; Oscar Lewis and Carroll D. Hall, *Bonanza Inn: America's First Luxury Hotel* (New York: Knopf, 1939), 67.

44. *SFC* (Jan. 27, 1888), 8.

45. *SFC* (May 19, 1919), 11.

46. *The Physiology of Taste: Harder's Book of Practical American Cookery* (San Francisco: no publisher, 1885), vi.

47. Lewis and Hall, *Bonanza Inn*, 74.

48. Herb Caen, *One Man's San Francisco* (New York: Doubleday, 1976), 229.

49. Elodie Hogan, "Hills and Corners of San Francisco," *Californian Illustrated Magazine* 5 (1893), 67.

50. Francine Brevetti, *The Fabulous Fior: Over 100 Years in an Italian Kitchen, The Fior d'Italia of San Francisco* (San Francisco: San Francisco Bay Books, 2005), 11, 17.

51. Helen Throop Purdy, *San Francisco: As it Was, As it Is, and How to See it* (San Francisco: P. Elder, 1912), 151; Edwords, *Bohemian San Francisco*, 100–104.

52. *SF-Call* (Oct. 20, 1909), 20.

53. Virgilio Luciani, cited by Brevetti, *The Fabulous Fior*, 32.

54. *SFC* (Jan. 19, 1921), A39.

55. Brevetti, *The Fabulous Fior*, 38–39.

56. Gumina, "Provincial Italian Cuisines"; Muscatine, *A Cook's Tour*, 257–59.

57. Brevetti, *The Fabulous Fior*, 75.

58. Brevetti, *The Fabulous Fior*, 71.

59. Brevetti, *The Fabulous Fior*, 78.

60. Edwords, *Bohemian San Francisco*, 48–49.

61. "T. Max Kniesche: oral history conducted in 1976 by Ruth Teiser," Regional Oral History Office, The Bancroft Library, University of California, Berkeley, 88.

62. *SFC* (Jan. 19, 1921), A39.

63. *SFC* (Oct. 26, 1922), 9.

64. Jessica Ellen Sewell, *Women and the Everyday City: Public Space in San Francisco, 1890–1915* (Minneapolis: University of Minnesota Press, 2011), 78–80.

65. Almira Bailey, *Vignettes of San Francisco* (San Francisco: San Francisco Journal, 1921), 25–26.

66. Herb Caen, introduction to *Frankly Speaking: Trader Vic's Own Story* (New York: Doubleday, 1973), xviii.

67. *The Coast: A Magazine of Western Writing* (Federal Writers' Project of the WPA, 1937), 4.

68. Quotes cited in Amy Reddinger, "Pineapple Glaze and Backyard Luaus: Cold War Cookbooks and the Fiftieth State," in *Pressing the Fight: Print, Propaganda, and the Cold War*, ed. Catherine Turner (Amherst: University of Massachusetts Press, 2012), 208.

69. "They Eat Well in San Francisco," *Harper's Bazaar* (February 1941), 36; "Chinese Oven," *Sunset Magazine* 91:6 (December 1943), 44–47; *Life* (Sept. 4, 1944), 79.

70. Herb Caen, *Don't Call It Frisco* (Garden City, NY: Doubleday, 1953), 276.

71. Herb Caen, *Guide to San Francisco* (Garden City, NY: Doubleday, 1957), 100; Lois Dwan, "Trader Vic's: Divided, not Conquered," *Los Angeles Times* (Sept. 7, 1980), 98.

72. Muscatine, *A Cook's Tour*, 308; Laresh Krishna Jayasanker, "Sameness in Diversity: Food Culture and Globalization in the San Francisco Bay Area and America, 1965–2005" (PhD diss., University of Texas, Austin, 2008), 210–12.

73. Muscatine, *A Cook's Tour*, 309.

74. Jayasanker, "Sameness in Diversity," 217–19.

75. Janet Fletcher, "Cappuccino by the Bay," *NYT* (July 16, 1989), XX19.

76. "Cecilia Chiang: An Oral History conducted in 2005–2006 by Victor Geraci, PhD," Regional Oral History Office, The Bancroft Library, University of California, Berkeley, 28, 45, 57–58.

77. Cecilia Chiang with Lisa Weiss, *The Seventh Daughter: My Culinary Journey from Beijing to San Francisco* (Berkeley: Ten Speed, 2007), 11–12; 50–51.

78. "Cecilia Chiang: An Oral History," 71.

79. "Cecilia Chiang: An Oral History," 77, 117–20.

80. David Kamp, *The United States of Arugula: How We Became a Gourmet Nation* (New York: Random House, 2006), 137, 142.

81. See Vanity Fair article on Chez Panisse (Oct. 2006), http://www.vanityfair.com/culture/features/2006/10/kamp_excerpt200610.

82. Kamp, *The United States of Arugula*, 150–57.

83. Thomas McNamee, *Alice Waters and Chez Panisse* (New York: Penguin, 2007), 109–11, 123; "All the Food was of Native Growth: Mrs. Touchard's California Dinner," *SFC* (Oct. 17, 1895), 5.

84. Kamp, *The United States of Arugula*, 92, 123, 129.

Chapter Six: San Francisco Cookbooks

1. *California Recipe Book* (San Francisco: Bruce's Printing House, 1872), 14.

2. *How to Keep a Husband or Culinary Tactics* (San Francisco: Cubery & Co., 1872), 9.

3. *What Mrs. Fisher Knows About Old Southern Cooking, Soups, Pickles, Preserves, Etc.* (San Francisco: Women's Co-Operative Printing Office, 1881), 3.

4. Karen Hess, "What We Know about Mrs. Abby Fisher and Her Cooking," in *What Mrs. Fisher Knows About Southern Cooking* (Bedford, MA: Applewood Books, 1995), 75–90; Andrew Warnes, *Hunger Overcome? Food and Resistance in Twentieth-Century African American Literature* (Athens: University of Georgia Press, 2004), 33. Fisher was born in South Carolina, then lived in Mobile as an adult before leaving the South in the 1870s.

5. *SFC* (May 12, 1881), 1.

6. H. J. Clayton, *Clayton's Quaker Cook-Book* (San Francisco: Women's Co-Operative Printing Office, 1883), iii.

7. *SFC* (May 12, 1881), 1.

8. Dan Strehl, "California's Culinary Literature," in *Encarnación's Kitchen: Mexican Recipes From Nineteenth-Century California*, ed. and trans. Dan Strehl (Berkeley: University of California Press, 2003), 28.

9. Encarnación Pinedo, *El cocinero español: Obra que contiene mil recetas valiosas y utiles para cocinar con facilidad en diferentes estilos* [The Spanish Cook: A Work Containing a Thousand Valuable and Useful Recipes to Cook with Ease in Different Styles] (San Francisco: Imprenta de E. C. Hughes, 1898).

10. *The Refugees' Cook Book: 50 Recipes for 50 Cents* (San Francisco: S. F. Archives Reprint, 1979 [orig. 1906]), 1.

11. Frank Morton Todd, *The Story of the Exposition*, vol. 4 (New York: Putnam, 1921), 327.

12. *SFC* (Mar. 8, 1915), 6.

13. Joseph Charles Lehner, *World's Fair Menu and Recipe Book; A Collection of the Most Famous Menus Exhibited at the Panama-Pacific International Exposition* (San Francisco: Lehner and Sefert, 1915).

14. L. L. McLaren, *Pan-Pacific Cook Book: Savory Bits from the World's Fare* (San Francisco: Blair-Murdock, 1915).

15. Belle De Graf, *Mrs. De Graf's Cook Book* (San Francisco: H. S. Crocker, 1922), 171.

16. *The Collected Recipes of the Alcatraz Women's Club* (San Francisco: Golden Gate National Park Association, 1995 [orig. 1952]).

17. *Council Cook Book* (San Francisco Section of the National Council of Jewish Women, 1909); *Soup to Nuts: Cook Book for Epicures*, Emanu-El Sisterhood (San Francisco, 1937 [orig. 1931]); Betty Kalis, Ron Moskowitz, *Out of Our Kitchen Closets: San Francisco Gay Jewish Cooking* (Congregation Sha'ar Zahav, 1987).

18. *Recipes of San Francisco Russian Molokans* (San Francisco: Women's Circle, First Russian Christian Molokan Church, 1973); Margaret H. Koehler, *Recipes from the Russians of San Francisco* (Riverside, CT: Chatham Press, 1974)

19. Koehler, *Recipes from the Russians*, 83.

20. Koehler, *Recipes from the Russians*, 58.

21. Koehler, *Recipes from the Russians*, 30, 38.

22. Johnny Kan with Charles L. Leong, *Eight Immortal Flavors* (Berkeley: Howell-North, 1963), 12.

23. Cecilia Chiang with Allan Carr, *The Mandarin Way* (Boston: Little, Brown & Co., 1974), 69.

24. Cecilia Chiang with Lisa Weiss, *The Seventh Daughter: My Culinary Journey from Beijing to San Francisco* (Berkeley: Ten Speed, 2007), 234.

25. Victor Hirtzler, *The Hotel St. Francis Cookbook* (Chicago: Hotel Monthly Press, 1919), 313.

26. Brian St. Pierre and Mary Etta Moose, *The Flavor of North Beach* (San Francisco: Chronicle, 1981).

Chapter Seven: Signature Dishes

1. "A Cook's Tour of San Francisco," *The Rotarian* (April 1947), 52. This may be the first appearance of the word "cappuccino" in an American publication.

2. SFC (Feb. 18, 1963), 1; Kevin Starr, *Coast of Dreams* (New York: Vintage, 2006), 474.

3. Clarence Edgar Edwords, *Bohemian San Francisco, Its Restaurants and Their Most Famous Recipes* (San Francisco: P. Elder, 1914), 64; SFC (Nov. 26, 1916), 33; SFC (Nov. 26, 1918), 2; SFC (Dec. 23, 1918), 8; and SFC (Nov. 26, 1919), 3.

4. Linda Hull Larned, "Some New York Discoveries," *Good Housekeeping* 47:4 (October 1908), 452.

5. SFC (Dec. 21, 1910), 1.

6. SFC (Dec. 25, 1910), 27.

7. SFC (Dec. 13, 1910), 11.

8. SFC (Mar 20, 1921), F2.

9. SF-Call (Nov. 17, 1912), 14.

10. Andrew Coe, *Chop Suey: A Cultural History of Chinese Food in the United States* (New York: Oxford University Press, 2009), 154–61.

11. Coe, *Chop Suey*, 176–79; Frank W. Aitken and Edward Hilton, *A History of the Earthquake and Fire in San Francisco* (San Francisco: E. Hilton, 1906), 131; Edwords, *Bohemian San Francisco*, 55; "Chop Suey Hoax Exposed," *The Mixer & Server* (Nov. 15, 1912), 34, citing *New York Telegraph*.

12. L. L. McLaren, *Pan-Pacific Cook Book: Savory Bits from the World's Fare* (San Francisco: Blair-Murdock Co., 1915), 78.

13. SFC (Oct. 15, 1899), 12. Hagiwara was ousted in 1900 but invited back in 1907 after the garden fell into disrepair. In between, he ran a private tea garden across from Golden Gate Park at Eighth Avenue.

14. SFC (June 30, 1901), 34.

15. Leslie Lieber, "The Inside Story of Chinese Fortune Cookies," *Los Angeles Times* (June 7, 1959), 125. See also Jennifer 8 Lee, *The Fortune Cookie Chronicles: Adventures in the World of Chinese Food* (New York: Twelve, 2008).

16. SF-Call (Aug. 20, 1899), 31.

17. SF-Call (Oct. 6, 1901), 10.

18. Herb Caen, *Only in San Francisco* (Garden City, NY: Doubleday, 1960), 108.

19. Menu courtesy of San Francisco Public Library.

20. Herb Caen, *Baghdad By the Bay* (Garden City, NY: Doubleday, 1949), 118.

21. Edwords, *Bohemian San Francisco*, 131.

22. Andrew Beahrs, "Slush on the Mizzentops, Butter in the Hold: Food on American Clipper Ships," *Gastronomica* 12:4 (Winter 2012), 42; SFC (Jan. 1, 1905), 9; Elisha Smith Capron, *History of California* (Boston: Jewett, 1854), 143.

23. *Sacramento Daily Union* (May 26, 1888), 1; Charles W. Haskins, *The Argonauts of California* (New York: Fords, Howard & Hulbert, 1890), 80.

24. Victor Hirtzler, *Hotel St. Francis Cook Book* (Chicago: Hotel Monthly Press, 1919), 64. On the price of oysters, see Mitchell Postel, "A Lost Resource: Shellfish in San Francisco Bay," *California History* 67:1 (March 1988), 30.

25. William Bronson, "Secrets of Pisco Punch Revealed," *California Historical Quarterly* 52:3 (Fall 1973), 231.

26. Robert O'Brien, *This is San Francisco* (San Francisco: Whittlesey House, 1948), 39.

27. SFC (May 21, 1920), 4.

28. SFC (July 29, 1883), 1.

29. SFC (Sept. 19, 1909), 32

30. Herb Caen, *One Man's San Francisco* (Garden City: Doubleday, 1976), 155.

31. Tony Abou-Ganim et al., *The Modern Mixologist: Contemporary Classic Cocktails* (Chicago: Agate Publishing, 2010), 29, 108.

32. Cory Doctorow, *Little Brother* (New York: Macmillan, 2008), 186; Gustavo Arellano, *Taco USA: How Mexican Food Conquered America* (New York: Scribner, 2012), 141; Tara Duggan, "The Silver Torpedo," SFC (Apr. 29, 2001).

33. John Roemer, "Cylindrical God," *SF Weekly* (May 5, 1993).

34. "El Burrito—The All American Food Item From Mexico," SFC (Dec. 23, 1979).

35. Captanian P., *Mémoires d'une déportée arménienne* (Paris: Flinkowski, 1919), 42. The Kitchen Sisters, "Birth Of Rice-A-Roni: The Armenian-Italian Treat," accessed on January 20, 2012, at http://www.npr.org/2008/07/31/93067862/birth-of-rice-a-roni-the-armenian-italian-treat.

36. "The DeDomenico Family: Growth of the Golden Grain Company Through Innovation and Entrepreneurship, an oral history conducted 1987–1989," Regional Oral History Office, The Bancroft Library, University of California, Berkeley, 22.

37. Jack Kerouac, *On the Road* (New York: Viking, 1957), 176.

38. Walter Colton, *The Land of Gold or: Three years in California [1846–1849]* (New York: D. W. Evans, 1860), 298.

39. "B.T. Babbitt's Pure Chemicals," *Sacramento Daily Union* (Mar. 26, 1860), 2.

40. SFC (Jan. 31, 1898), 18. See also Ruth Allman, *Alaska Sourdough* (Anchorage: Alaska Northwest, 1976); and Jack London, *White Fang* (New York: Macmillan, 1906), 152.

41. James Tyson, *Diary of a Physician in California* (New York: D. Appleton, 1850), 9; *Sacramento Daily Union* (Dec. 14, 1860), 4.

42. SFC (Mar. 6, 1904), 7.

43. Frank Morton Todd, *The Story of the Exposition*, vol. 4 (New York: Putman, 1921), 294; Georges Lanson, *Guide des Francais en California* (San Francisco: G. Lanson, 1917), xxix.

44. Langley's San Francisco Directories have a Boudin bakery on Dupont in 1852; then Union, just off Dupont; then Green, just off Dupont; then back on Dupont by 1876.

45. *Sunset Kitchen Cabinet Recipes* (San Francisco: Lane Publishing, 1944), 99.

46. Leo Kline and T. F. Sugihara, "Microorganisms of the San Francisco Sour Dough Bread Process," Applied Microbiology 21:3 (March 1971), 456–58. For more on the difference between sourdough techniques and yeasted breads, see William Rubel, *Bread: A Global History* (London: Reaktion Books, 2011).

~

Bibliography

Abbreviations

DAC: *Daily Alta California*
NYT: *New York Times*
SFC: *San Francisco Chronicle*
SF-Call: *San-Francisco Call*

Bibliography

Anderson, M. Kat. *Tending the Wild: Native American Knowledge and the Management of California's Natural Resources.* Berkeley: University of California Press, 2006.

Arellano, Gustavo. *Taco USA: How Mexican Food Conquered America.* New York: Scribner, 2012.

Bancroft, Hubert Howe. "Mongolianism in America." *Essays and Miscellany.* San Francisco: The History Company, 1890.

Beahrs, Andrew. "Slush on the Mizzentops, Butter in the Hold: Food on American Clipper Ships." *Gastronomica* 12:4 (Winter 2012), 37–45.

Bean, Lowell John. *The Ohlone, Past and Present: Native Americans of the San Francisco Bay Region.* Menlo Park, CA: Ballena Press, 1994.

Bergeron, Victor. *Frankly Speaking: Trader Vic's Own Story.* New York: Doubleday, 1973.

Booker, Matthew. *Down By The Bay: San Francisco's History Between the Tides.* Berkeley: University of California Press, 2013.

Boyd, Nan Alamilla. *Wide-Open Town: A History of Queer San Francisco to 1965.* Berkeley: University of California Press, 2003.

Brechin, Gary. *Imperial San Francisco: Urban Power, Earthly Ruin*. Berkeley: University of California Press, 2006.

Brevetti, Francine. *The Fabulous Fior: Over 100 Years in an Italian Kitchen, The Fior d'Italia of San Francisco*. San Francisco: San Francisco Bay Books, 2005.

Briscoe, John. *The Tadich Grill: The Story of San Francisco's Oldest Restaurant, with Recipes*. Berkeley, CA: Ten Speed Press, 2002.

Brook, James, Chris Carlsson, and Nancy J. Peters, eds. *Reclaiming San Francisco: History, Politics, Culture*. San Francisco: City Lights, 2001.

Brucato, John G. *The Farmer Goes to Town: The Story of San Francisco's Farmer's Market*. San Francisco: Burke Publishing, 1948.

Carlsson, Chris, ed. *Ten Years That Shook the City: San Francisco, 1968–1978*. San Francisco: City Lights, 2011.

Chiang, Cecilia, with Lisa Weiss. *The Seventh Daughter: My Culinary Journey from Beijing to San Francisco*. Berkeley: Ten Speed Press, 2007.

Cinotto, Simone. *Soft Soil, Black Grapes: The Birth of Italian Winemaking in California*. New York: New York University Press, 2012.

Clayton, H. J. *Clayton's Quaker Cook Book*. San Francisco: Women's Co-Operative Printing Office, 1883.

Cobble, Dorothy Sue. *Dishing It Out: Waitresses and Their Unions in the Twentieth Century*. Urbana: University of Illinois Press, 1991.

Coe, Andrew. *Chop Suey: A Cultural History of Chinese Food in the United States*. New York: Oxford University Press, 2009.

Conlin, Joseph Robert. *Bacon, Beans, and Galantines: Food and Foodways on the Western Mining Frontier*. Reno: University of Nevada Press, 1987.

De Graaf, Lawrence B., Kevin Mulroy, and Quintard Taylor, eds. *Seeking El Dorado: African Americans in California*. Los Angeles: Autry Museum of Western Heritage, 2001.

De Talavera Berger, Frances, and John Parke Custis. *Sumptuous Dining in Gaslight San Francisco 1875–1915*. Garden City, NY: Doubleday, 1985.

Dillon, Richard. *North Beach: The Italian Heart of San Francisco*. Novato, CA: Presidio, 1985.

Diner, Hasia R. *Hungering for America: Italian, Irish, and Jewish Foodways in the Age of Migration*. Cambridge, MA: Harvard University Press, 2001.

Dubin, Margaret Denise, and Sara-Larus Tolley. *Seaweed, Salmon, and Manzanita Cider: A California Indian Feast*. Berkeley, CA: Heyday, 2008.

Edwords, Clarence Edgar. *Bohemian San Francisco, Its Restaurants and Their Most Famous Recipes*. San Francisco: P. Elder, 1914.

Fairfax, Sally K., et al. *California Cuisine and Just Food*. Cambridge, MA: MIT Press, 2012.

Fisher, Abby. *What Mrs. Fisher Knows About Old Southern Cooking, Soups, Pickles, Preserves, Etc.* San Francisco: Women's Co-Operative Printing Office, 1881.

Fleming, Arthur. *My Secret San Francisco: How to Eat, Drink and Swing in San Francisco on Almost No Money*. San Francisco: Sea Classics Press, 1966.

Garvey, John, and Karen Hanning. *Irish San Francisco*. San Francisco: Arcadia Publishing, 2008.

Gribben, Arthur, ed. *The Great Famine and the Irish Diaspora in America*. Boston: University of Massachusetts Press, 1999.

Gumina, Deanna Paoli. *The Italians of San Francisco, 1850–1930*. New York: Center for Migration Studies, 1978.

Gumina, Deanna Paoli. "Provincial Italian Cuisines: San Francisco Conserves Italian Heritage." *The Argonaut: Journal of the San Francisco Historical Society* 1:1 (Spring 1990).

Habal, Estella. *San Francisco's International Hotel: Mobilizing the Filipino American Community in the Anti-Eviction Movement*. Philadelphia: Temple University Press, 2007.

Hess, Karen. "What We Know about Mrs. Abby Fisher and Her Cooking." *What Mrs. Fisher Knows About Southern Cooking*. Bedford, MA: Applewood Books, 1995.

Issel, William, and Robert W. Cherny. *San Francisco, 1865–1932: Politics, Power, and Urban Development*. Berkeley: University of California Press, 1986.

Jacknis, Ira. *Food in California Indian Culture*. Berkeley: University of California Press, 2004.

Jayasanker, Laresh Krishna. "Sameness in Diversity: Food Culture and Globalization in the San Francisco Bay Area and America, 1965–2005." PhD diss., University of Texas, Austin, 2008.

Jones, Terry L., and Kathryn A. Klar. *California Prehistory: Colonization, Culture, and Complexity*. Lanham, MD: AltaMira Press, 2010.

Kamp, David. *The United States of Arugula: How We Became a Gourmet Nation*. New York: Random House, 2006.

Kroeber, Alfred Louis, et al. *A Mission Record of the California Indians*, vol. 8. Berkeley: The University Press, 1910.

Laguerre, Michel. *The Global Ethnopolis: Chinatown, Japantown and Manilatown in American Society*. New York: St. Martin's Press, 2000.

Lemke-Santangelo, Gretchen. *Abiding Courage: African American Migrant Women and the East Bay Community*. Chapel Hill: University of North Carolina Press, 1996.

Levinson, Marc. *The Great A&P and the Struggle for Small Business in America*. New York: Hill & Wang, 2011.

Lévy, Daniel. *Les Français en Californie*. San Francisco: Grégoire, Tauzy, 1885.

Levy, Harriet Lane. *920 O'Farrell Street: A Jewish Girlhood in Old San Francisco*. Berkeley: Heyday Books, 1996 [orig. 1937].

Lewis, Oscar, and Carroll D. Hall. *Bonanza Inn: America's First Luxury Hotel*. New York: Knopf, 1939.

Lightfoot, Kent G. *Indians, Missionaries, and Merchants: The Legacy of Colonial Encounters on the California Frontiers*. Berkeley: University of California Press, 2006.

Lightfoot, Kent, and Otis Parrish. *California Indians and their Environment*. Berkeley: University of California Press, 2009.

Ling, Huping. *Surviving on the Gold Mountain: A History of Chinese American Women and Their Lives*. Albany: SUNY Press, 1998.

Lobo, Susan. *Urban Voices: The Bay Area American Indian Community*. Tucson: University of Arizona Press, 2002.

Lord, Jack, and Jenn Shaw. *Where to Sin in San Francisco*. San Francisco: Guggenheim, 1939.

Margolin, Malcolm. *The Ohlone Way: Indian Life in the San Francisco-Monterey Bay Area*. Berkeley: Heyday Books, 1978.

McClintock, Nathan. *From Industrial Garden to Food Desert: Unearthing the Root Structure of Urban Agriculture in Oakland, California*. Berkeley: University of California Press, 2008.

McNamee, Thomas. *Alice Waters and Chez Panisse*. New York: Penguin, 2007.

Minnich, Richard A. *California's Fading Wildflowers: Lost Legacy and Biological Invasions*. Berkeley: University of California Press, 2008.

Moore, Shirley Ann Wilson. "'Not in Somebody's Kitchen': African American Women Workers in Richmond, California, and the Impact of World War II." *Writing the Range: Race, Class, and Culture in the Women's West*. Edited by Elizabeth Jameson and Susan Armitage. Norman: University of Oklahoma Press, 1997.

Moore, Shirley Ann Wilson. "'Your Life Is Really Not Just Your Own': African American Women in Twentieth Century California." *Seeking El Dorado: African Americans in California*. Edited by Lawrence B. De Graaf et al. Los Angeles: Autry Museum of Western Heritage, 2001.

Muscatine, Doris. *A Cook's Tour of San Francisco: The Best Restaurants and Their Recipes*. New York: Scribner, 1963.

Newell, Quincy D. *Constructing Lives at Mission San Francisco: Native Californians and Hispanic Colonists, 1776–1821*. Albuquerque: University of New Mexico, 2009.

O'Connell, Daniel. *The Inner Man: Good Things to Eat and Drink and Where to Get Them*. San Francisco: Bancroft, 1891.

Okamoto, Ariel Rubissow, and Kathleen M. Wong. *Natural History of San Francisco Bay*. Berkeley: University of California Press, 2011.

Praetzellis, Mary, and Adrian Praetzellis. "'Black Is Beautiful': From Porters to Panthers in West Oakland." Report by the Anthropological Studies Center at Sonoma State University (June 2004).

Purdy, Helen Throop. *San Francisco: As it Was, As It Is, and How to See It*. San Francisco: P. Elder, 1912.

Roberts, J. A. G. *China to Chinatown: Chinese Food in the West*. London: Reaktion Books, 2002.

Rosenbaum, Fred. *Cosmopolitans: A Social and Cultural History of the Jews of the San Francisco Bay Area*. Berkeley: University of California Press, 2009.

Rubel, William. *Bread: A Global History*. London: Reaktion Books, 2011.

San Francisco: The Bay and Its Cities. Federal Writers' Project Staff of the Works Progress Administration for Northern California, 1940.

Sánchez, Rosaura. *Telling Identities: The Californio Testimonios*. Minneapolis: University of Minnesota Press, 1995.

Scott, Mel. *The San Francisco Bay Area: A Metropolis in Perspective*. Berkeley: University of California Press, 1985.

Sewell, Jessica Ellen. *Women and the Everyday City: Public Space in San Francisco, 1890–1915*. Minneapolis: University of Minnesota Press, 2011.

Sinclair, Mick. *San Francisco: A Cultural and Literary History*. Oxford: Signal Books, 2004.

Smith, James R. *San Francisco's Lost Landmarks*. San Francisco: Quill Driver, 2005.

Snedden, Genevra Sisson. *Docas, the Indian Boy of Santa Clara*. Boston: D. C. Heath, 1899.

Solnit, Rebecca. *Infinite City: A San Francisco Atlas*. Berkeley: University of California Press, 2010.

Sparks, Edith. *Capital Intentions: Female Proprietors in San Francisco, 1850–1920*. Chapel Hill: University of North Carolina Press, 2006.

St. Pierre, Brian, and Mary Etta Moose. *The Flavor of North Beach*. San Francisco: Chronicle, 1981.

Starr, Kevin. *California: A History*. New York: Modern Library, 2005.

Street, Richard Steven. *Beasts of the Field: A Narrative History of California Farmworkers, 1769–1913*. Stanford, CA: Stanford University Press, 2004.

Strehl, Dan, ed. and trans. *Encarnación's Kitchen: Mexican Recipes from Nineteenth-Century California*. Berkeley: University of California Press, 2003.

Strehl, Dan. *One Hundred Books on California Food and Wine*. Los Angeles: Book Collectors of Los Angeles, 1990.

Stromberg, Mark R., Jeffrey D. Corbin, and Carla Marie D'Antonio. *California Grasslands: Ecology and Management*. Berkeley: University of California Press, 2007.

Thompson, Mark. *Vintage California Cuisine: 300 Recipes from the First Cookbooks Published in the Golden State*. Philadelphia: Seasonal Chef Press, 2012.

Thompson, Ruth, and Louis Hanges. *Eating Around San Francisco*. San Francisco: Suttonhouse, 1937.

Voss, Barbara L. *The Archaeology of Ethnogenesis: Race and Sexuality in Colonial San Francisco*. Berkeley: University of California Press, 2008.

Walther, Rudolph J. *A Tour to San Francisco, California and Return*. Philadelphia: Walther Printing House, 1916.

Wells, Evelyn. *Champagne Days of San Francisco*. New York: D. Appleton, 1939.

Wollenberg, Charles. *Golden Gate Metropolis: Perspectives on Bay Area History*. Berkeley: University of California Press, 1985.

Yee, Alfred. *Shopping at Giant Foods: Chinese American Supermarkets in Northern California*. Seattle: University of Washington Press, 2003.

Yung, Judy. *Unbound Voices: A Documentary History of Chinese Women in San Francisco*. Berkeley: University of California Press, 1999.

Index

~

About the Author

Erica J. Peters is cofounder and director of the Culinary Historians of Northern California and the author of *Appetites and Aspirations in Vietnam: Food and Drink in the Long Nineteenth Century* (2012).